"Both governance and sustainable development are complex and often poorly understood concepts. However, both are essential to understanding how we can hope to address the current planetary crisis. Meri Juntti's book is an invaluable aid to this task: unpicking these multi-faceted and contested concepts and then applying them to the pressing issues of climate change and global food security. In doing so, complex questions and arguments around sustainable governance are presented here in an accessible, though authoritative, manner with the judicious use of effective learning tools to aid the reader, be they a student or a concerned citizen".

Peter Hough, *Professor at Middlesex University, London, UK*

"Drawing on years of excellent scholarship and empirical experience 'on the ground', Meri Juntti's *Governing for Sustainable Development: 21st Century Challenges and Principles* makes essential reading for undergraduate, postgraduate students, and scholars on governance and sustainability. With an excellent combination of breadth and depth, using contemporary examples, Juntti dextrally illustrates the complexity of current governance global challenges. She does so using accessible language and approach that will make this a key reference in higher education".

Julia Martin-Ortega, *Professor and Associate Director of Sustainability Research Institute at University of Leeds, UK*

"Understanding and acting on the many-dimensioned challenges – environmental, social, cultural, political, economic, institutional – we face can feel like a daunting task, especially in the context of multiple crises. Meri Juntti has given us an important and accessible guide to navigating these difficult paths. By examining challenges, debates, and principles for analysing governance, she points the way towards improved governance for sustainable development through better analysis and action".

John Turnpenny, *Associate Professor, School of Politics, Philosophy and Area Studies, University of East Anglia, UK*

Governing for Sustainable Development

This textbook explores the meaning of good governance and sustainable development in the present-day context of a growing planetary crisis, socio-economic polarisation, and politicisation of knowledge.

Part 1 sets the stage by exploring theories of power, network governance, and global governance. Meri Juntti contrasts prevalent and emerging framings to arrive at a definition of sustainable development as a transformation encompassing the economic system, societal values, and the relationship of humanity with nature, and identifies distinct approaches and solutions for advancing this. Part 2 delves into the twin global issues of climate change and food security. Their drivers and manifestations are examined in detail and a range of governance responses – from global strategies and policies through to local-level case studies – are explored in the light of earlier theorisations. Part 3 focuses on lessons for good governance and sets out five key challenges that governance for sustainable development needs to negotiate, as well as the four principles that are central to its success. A concluding chapter considers what good governance for sustainable development could mean in practice.

Balancing theory with real-world examples, and featuring learning outcomes, further reading, and links to relevant media content, *Governing for Sustainable Development* will be a highly valuable resource for advanced undergraduate and postgraduate students of sustainable development, geography, public policy, politics, and international relations.

Meri Juntti, PhD, SFHEA, is Associate Professor of Environmental Governance in the School of Law and Social Sciences at Middlesex University, UK. She has co-designed and presently leads an MA Programme (MA Global Governance and Sustainable Development), co-leads the Sustainable Development Research Cluster, and chairs the University's Sustainability Steering Group. She has undertaken a number of externally funded research projects in the field of environmental governance and has published several single and co-authored publications in this area. Dr Juntti holds an honorary position at the Bartlett School of Planning, UCL, and is a Trustee of the London Development Trust.

Governing for Sustainable Development

21st Century Challenges and Principles

Meri Juntti

Routledge
Taylor & Francis Group
LONDON AND NEW YORK
earthscan
from Routledge

Designed cover image: Getty Images

First published 2027
by Routledge
4 Park Square, Milton Park, Abingdon, Oxon OX14 4RN

and by Routledge
605 Third Avenue, New York, NY 10158

Routledge is an imprint of the Taylor & Francis Group, an informa business

British Library Cataloguing-in-Publication Data
A catalogue record for this book is available from the British Library

ISBN: 9781032005997 (hbk)
ISBN: 9781032015880 (pbk)
ISBN: 9781003179207 (ebk)

DOI: 10.4324/9781003179207

Typeset in Sabon
by Deanta Global Publishing Services, Chennai, India

Contents

List of Figures *vi*

1 Introduction 1

Part 1
Setting the scene – governing in the age of crises **15**

2 The 21st-century governance context 17
3 Sustainable development as transformation 36
4 Governance and power 59

Part 2
Understanding global governance issues and solutions **81**

5 Governing the climate crisis 83
6 Governing for food security 104

Part 3
Lessons for good governance for sustainable development **125**

7 Identifying good governance for sustainable development 127
8 Conclusions 150

Index *165*

Figures

1.1 The doughnut model of sustainable development (adapted from: Raworth, 2017) 4
2.1 Share of global aggregate income or wealth across three income groups: the bottom 50%, the middle 40%, and the top 10%. (adapted from Chancel et al. 2026 CC BY-NC-SA 4.0 Data from: wir2026.wid.world/methodology) 19
2.2 The circular flow diagram depicting the economic system where capital and resources are combined to produce goods and services in exchange for labour. (adapted from Raworth2017 Copyright © 2017 Kate Raworth.). 25
2.3 Monthly incomes across regions in 2025. (adapted from Chancel et al. 2026. CC BY-NC-SA 4.0. Data from wir2026.wid.world/methodology). 28
3.1 The Earth and the nine Earth system boundaries, six of which are now in the zone of high risk of unprecedented developments. (Adapted from Stockholm Resilience Centre, 2025. Licensed under CC BY-NC-ND 3.0. Credit: "Azote for Stockholm Resilience Centre, based on analysis in (Sakschewski et al., 2025)". 40
3.2 The Millennium Ecosystem Assessment classification of ecosystem services and links to the constituents of wellbeing. (adapted from MEABoard 2005: 7) 42
3.3 The restorative continuum and the four major categories of activities pertaining to the restoration of nature and its ecosystems and biodiversity. (adapted from (Gann et al., 2019)). 44
3.4 Relative decoupling of resource extraction from economic growth (GDP line) happens when GDP grows faster than resource use. Absolute decoupling, on the other hand, implies a negative correlation between resource extraction and GDP. 49
3.5 The doughnut representing a safe and just space for humanity. (adapted from:Raworth 2017) 51

5.1 Annual CO_2 emissions from fossil fuels and industry since 1750 (Ritchie, H., 2026). (Data Source: Global Carbon Budget (2025) – with major processing by Our World in Data). 86
5.2 Growth rates in total energy demand, GDP and electricity demand, and the growth share of different energy sources in 2024 (adapted from IEA2025 under the IEA CC BY 4.0 licence). 87
6.1 The prevalence of food insecurity globally across regions over the past 10 years to 2025, demonstrating the impact of the Covid-19 pandemic in 2020. Oceania is not displayed due to lack of sufficient data. 107
7.1 The governance cycle. 138
7.2 The intervention theory underpinning the key governance intervention aiming to mitigate the pollution emitted by Finnish pulp and paper mills. The intervention theory describes how the intervention, a permit-based system whereby each mill needs to first apply for and then regularly renew a permit for operation, is intended to work (Mickwitz, 2003b). 142

1 Introduction

1.1 Aims and scope of the book

This book, *Governing for Sustainable Development* offers a critical exploration of governance and what might be termed good governance for sustainable development in the 21st century. Its main aim is to help the reader build an understanding of some of the central challenges that stakeholders such as communities, and governmental, and non-governmental institutions face in influencing change for a sustainable future and to provide a critical understanding of good governance and sustainability itself. It sets the scene with two key components: an examination of the present-day governance context of an interdependent, globalised world that is moving into the new uncharted era of the Anthropocene (Crutzen & Stoermer, 2000; Folke et al., 2021), and a critical examination of the term sustainable development (WCED, 1987; Lélé, 1991; Redclift, 2005; Kothari et al., 2014; Ruggerio, 2021). The book covers a range of approaches to governance at the scale of international institutions, nation states, regional networks, and localities, and provides a critical perspective by engaging with theorisations of social phenomena such as crisis, power, justice, reflexivity, and sustainability. Two issue-based chapters engage with the key global challenges of climate change and food security and provide examples and a more detailed understanding of existing governance instruments and challenges.

The book is based on three key premises. Firstly, the acknowledgement that the dominant organising force today is the global economy, a system of extractive, neoliberal capitalism (Piketty, 2014). This system positions economic growth as the principal metric of development (Jackson, 2017). While the 20th century mastered the mechanised extraction of value from labour and natural resources, in the 21st century, growth is increasingly leveraged by innovation and complex technologies (Castells, 2009). Technologies focused on information and data transfer and processing are now ubiquitous. Algorithms and Artificial Intelligence (AI) permeate not only science and medicine but our everyday lives, modulating how and what information we access, how we make decisions about production and consumption practices, and even how we establish relationships. These technologies are primarily harnessed to serve the growth imperative through efficiency gains in

DOI: 10.4324/9781003179207-1

value extraction, integrating the planet ever more comprehensively into the logic of the growth economy. This is visible in the relentless commodification and digitisation of our lives and living environments.

Secondly, we exist in an increasingly destabilised planetary system. Gradual human-induced changes to the environment throughout the industrial era have resulted in unprecedented risks (Beck, 2016; IPCC, 2007). These can be understood in terms of so-called 'Earth system boundaries' (Steffen et al., 2015) coined in 2015 by researchers in the Stockholm Resilience Centre. The Earth system boundaries are the nine crucial environmental challenges faced by humanity in the 21st century. They mark the extent of change that planet Earth can tolerate; for example, in climate, land cover, biodiversity, and chemical loading without becoming significantly destabilised (Folke et al., 2021). Science shows that at the time of writing, six of the nine boundaries have been breached, with potentially devastating impacts on planetary equilibrium affecting the current premises of life on Earth (Richardson et al., 2023). Some suggest that this marks a departure from the geological era of the Holocene during which human civilisations became established and we were able to reach current levels of development and wellbeing, and that this new era that we now inhabit should be named the Anthropocene as it presents us with new human-induced circumstances and risks (Crutzen & Stoermer, 2000). These developments have inspired a rhetoric of multiple crises (such as climate crisis and extinction crisis) among scientists and the public (Grossman, 2020; HM Government, 2019; Oxfam, 2023).

Finally, a third premise is that globally, we are subject to the opposite forces of fragmentation and interdependence. While the global balance of powers is undergoing an evolution and a backlash to economic globalisation is evident in the 'Trump 2' era, we remain interconnected through digital flows of information, ideas and finance, the flow of goods through international commodity networks and the migration of people across the globe (KOF Swiss Economic Institute, 2024; Potrafke, 2015; Weiss & Wilkinson, 2018). Alongside this, and particularly at the local scale, inequality and socio-economic polarisation are persistently increasing (Piketty, 2014). The first decades of the 21st century have seen growing precarity of livelihoods and increasing fragmentation of society along socio-economic and political divides (WEF, 2025). Internationally, geo-political tensions and the risk of state conflict are high and visible in the many tragic wars that have marked the 2020s.

The above complexity means that in the 21st century, governance, understood broadly as goal-oriented action aimed at social coordination in the public domain (Bevir, 2011), takes place in a context of a multiplicity of risks and uncertainties, complex interdependencies, plural values and knowledge claims and conflicts of interest with associated winners and losers (Beck, 1992). Now more than ever governance for sustainable development requires a detailed appreciation of our complex socio-technical systems where social, political-institutional, economic, technological, and ecological dimensions are interrelated (Ruggerio, 2021; Voulvoulis et al., 2022). This calls for systems thinking, an integrated approach to understanding complex problems by examining the interconnections and interdependencies between different components of a system. The recent global Covid pandemic and frequent disastrous weather events ranging from deadly storms to fires and floods exemplify a range of risks with complex causes and impacts and emphasise the need for attention across multiple scales from the local to the planetary. Successful governance interventions in these challenging circumstances require careful consideration of governance goals (how to define the common good), approaches (what is the best way to achieve sustainable solutions in this context) and practices (how to ensure that we have everybody on board).

1.2 Sustainable development

In 1987, the Brundtland Report defined sustainable development as 'development that meets the needs of the present, without compromising the ability of future generations to meet their own needs' (WCED, 1987). This definition centred the notion of needs and stipulated a timeframe that encompasses future generations. Those remain two key tenets of sustainable development. But the concept has been compromised by questions concerning the nature of human needs, how exactly their fulfilment can be guaranteed for all and the reliability of future impact assessments. Furthermore, some suggest that sustainable development is a Western philosophy that does not address the inequalities that date back to the colonial era. This ambiguity has been lamented by those who are keen to achieve a more equitable and sustainable future and exploited by many who benefit from the current situation.

In 2015, the Agenda 2030 outlined 17 United Nations Sustainable Development Goals (UN SDGs), providing a globally relevant framework addressing many of the above ambiguities (UN General Assembly, 2015). The 17 goals encompass basic needs such as access to food, health, and education (goals 1–4); access to services, goods, livelihoods, and healthy living environments (goals 5–12); environmental sustainability (goals 13–15); and responsible governance based on fair and equal partnerships (goals 16–17). With their specific sub-targets and indicators, the SDGs aim to address the lack of progress in sustainable development by clarifying what exactly it means in practice and providing universal metrics for assessing progress.

But the SDGs have been criticised for not addressing the deeper drivers of unsustainability and inequality embedded in the global economic system and in particular, the imperative of extractive economic growth that it sets (Jackson, 2017; Robra & Heikkurinen, 2021). In her radical thesis for an alternative model of development, Kate Raworth (2017) suggests that instead of fixating on perpetual economic growth, narratives of development should aim for a state of equilibrium, above the baseline of fulfilling basic social needs and below the ceiling of ecological sustainability (Figure 1.1). Various emerging decolonial, degrowth and postgrowth approaches encourage

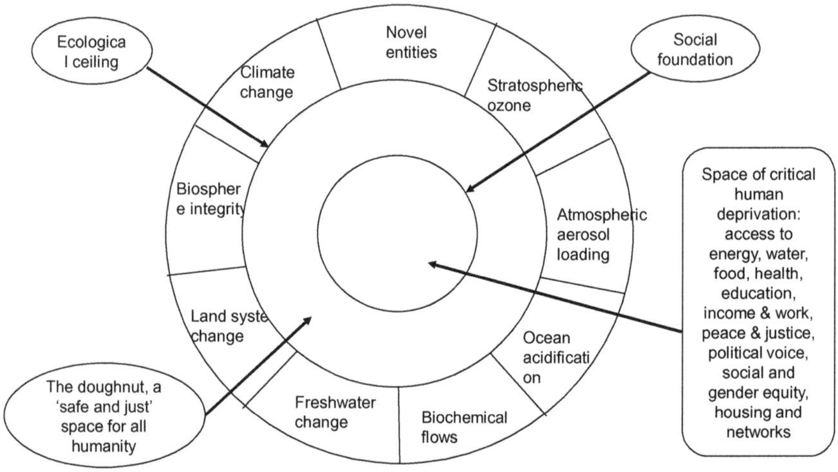

Figure 1.1 The doughnut model of sustainable development (adapted from: Raworth, 2017)

thinking outside the dominant economic values and institutions that uphold them to provide space for solutions that focus on genuine development for all and avoid institutionalised value and power biases underpinning current problems (Kallis et al., 2020; Kothari et al., 2014).

In the context of persistently growing inequality and polarising societal values and norms, sustainable development hinges on a nuanced and critical understanding of what exactly shapes access to wellbeing and ability to thrive for individuals, communities, countries, and regions (Jackson, 2017). This understanding calls for wide recognition and representation of different perspectives in defining the goals and means of development and governance. Some argue that this recognition of equal right to be recognised extends beyond humans to other living and non-living entities constituting the planetary system. While mainstream approaches to sustainable development have begun to integrate notions of equity, these competing approaches centre on the notion of planetary justice (Gupta et al., 2023). The definition of sustainable development advanced in this book draws on these new critical conceptualisations of sustainability.

Environmental sustainability, or more precisely, the stability of the Earth system (Richardson et al., 2023; Steffen et al., 2015) remains a fundamental premise of sustainable development. The Earth system perspective means that the scope of governance is expanded from the global to the planetary. The implications will be fully discussed in Chapters 3 and 7, but in the main, the planetary focus acknowledges the inextricable interdependence of nature and humans in the context of the dynamic Earth system. An example of this is the susceptibility of food security (Chapter 6) to climate change (Chapter 5), a planetary scale process hinging on atmospheric regulation (see Chapter

6). The planetary scope of governance therefore decentres the human perspective and draws into equal focus non-human entities such as forests that play a role in climate regulation, as meriting consideration in terms of needs, rights, and the ability to thrive (San Martín & Wood, 2021). Governing for sustainable development becomes more than just managing human activities for the good of human societies.

Chapter 3 delves deeper into this critical discussion of sustainable development and sets out the definition used in this book in more detail, as a *fundamental societal transformation towards equitable and regenerative relations among people and the planet, addressing the core drivers of unsustainability*.

1.3 Governance

If sustainable development has been deemed a nebulous term, the same can be said for governance. For the purposes of this book, governance is not limited to the actions of governments or formal governance actors whose remits are set out in legislation. Individuals, businesses, and non-governmental organisations can be powerful governance actors. Chapter 3 will argue that governance should also recognise the agency of non-humans such as forests, water bodies, and technology (Clark & Szerszynski, 2020). Governance is defined in multiple ways by different authors and practitioners, and while Bevir's (2011) definition of 'goal-oriented action regarding social coordination in the public domain' covers most of these, this section discusses a number of definitions intended to help the reader form a more solid understanding of what is at stake in governance.

The Cambridge Dictionary defines 'governance' as 'the way that organizations or countries are managed at the highest level, and the systems for doing this'. This is a narrow definition implying a hierarchical set-up and focusing only on the top of the hierarchy. Pierre and Peters (2021, p. 2) stay with this top-down focus but draw attention to governance as a dynamic process whereby 'governments, along with their private sector allies, steer the economy and society toward collective goals'. While this expands the scope of governance beyond the public sector, for many writers, and for the purposes of this book, governance is about more than state-society relations. In fact, many governance scholars state that such a dichotomy – a division into 'state' and 'society' – is misleading.

The notion of a governance 'shift' or 'turn' (Jordan et al., 2005; Kooiman, 2003) is helpful to begin to unpack the concept itself. Governance shift is a term used for the trends in the behaviour of national governments and governmental departments in public goods and services provision that first became evident in the 1990s. The rise of neoliberalism, a free-market ideology, threw into question the efficiency of the hierarchical top-down command-and-control approach to public regulation (Bevir, 2011). The public sector began to adopt management criteria from the private sector, seeking efficiency gains by slimming down operations, outsourcing public service provision to the private sector, and adopting so-called new economic policy instruments that relied on economic incentives instead of regulation to deliver behavioural change (Jordan et al., 2005). This integration of 'market logic' (demand-led provision, where price signals inform optimal volume of production) into the public sector led to the inclusion of private businesses and an increasing range of interest groups within the sphere of governance (Rhodes, 2007).

Public goods and services are central to both sustainable development and governance. They are services and goods that benefit many and are often crucial for the sovereignty of national states, but do not comply with the characteristics of private property, private means of production and property rights. Examples of such public goods at a global scale are public health, a stable climate, carbon sequestration (the capacity of forests, oceans, and soils to store carbon) and the internet (Brousseau et al., 2012). While the production and maintenance of private goods or common pool resources (semi-private goods such as public roads and street lighting for example) can be incentivised and enabled through regulating who uses them (only car users who have paid road tax for example), completely public goods are non-excludable, which means that people cannot be stopped from consuming them, and charging per use is often difficult (Ostrom et al., 1999). Markets and pricing alone cannot therefore be expected to manage their provision. The provision of public and semi-private goods needs concerted action by governments, communities, businesses, and even individual people (Kaul, 2012). This activity can be termed governance.

In this book, governance therefore refers to *the institutions and practices of decision making and intervention regarding issues of public interest, that is undertaken by formal and informal governance actors, often situated in networked relations of power* (Morrison et al., 2019; Sørensen, 2002). This definition of governance blurs the boundaries of the state and society, public and private, and people and nature. It refers to the processes and interactions through which all kinds of interests and actors combine to produce the instruments, practices and outcomes of governance interventions (Bevir, 2011; Rhodes, 2007; Weiss & Wilkinson, 2018). In line with the planetary focus associated with sustainable development, 'public interest' refers to human and non-human actors and entities and their ability to thrive. Instead of a hierarchical formation where 'power' sits with the top executive – the government, governance is increasingly viewed as 'polycentric' (Aligica & Tarko, 2012). This refers to a situation where multiple 'decision centres' (organisations, governmental bodies, interest groups etc.) operate autonomously but under an overarching set of rules to meet commonly agreed targets for example in providing public transport connections or in avoiding deforestation (Aligica & Tarko, 2012).

At the international scale, governance is often termed 'global governance', referring to the collaborations of state and non-state actors in advocating, establishing agreements about problem definitions, goals and actions addressing issues that have a global scope in terms of impact and/or causality (Sell, 2018). While at the national scale, governments often form the executive centre of a polycentric governance network, domestic institutions and contexts vary hugely and shape the agendas and actions of nation states in the global governance arena (Weiss, 2003). Therefore, global governance intertwines not only the conventionally separate spheres of the state and markets but also the vertical layers of local, regional, national, and international scales of action (Weiss, 2003). In the international arena, sovereign nation states are guided by the two fundamental goals of extending domestic economic activity into the international sphere and maintaining the international 'balance of power' (Polanyi, 1944).

'Good governance' is increasingly seen as central to delivering sustainable development and Chapters 4 and 7 provide approaches and principles for analysing governance. There are some criteria that are commonly held as key to good governance, such as transparency and accountability of governance institutions, decisions, and instruments (Weiss & Wilkinson, 2018). When it comes to sustainable development, an appreciation of complex systems and issues of justice and legitimacy is crucial, suggesting that governance goals also matter, not just the means of getting there. But more tangible guidance on what counts as good governance can best be understood in context. Due to the inherent high risks, some issues, such as the management of toxic waste, require a regulatory science-led approach (Hough, 2021). Other issues such as sustainable forest management, which involves the coordination of diverse actors and local knowledge, may be best addressed bottom-up, in a polycentric formation, taking the lead from communities who have co-existed with those habitats for centuries (Ham & Hill, 1993). A good governance initiative has legitimate aims and means and achieves relevant and intended outcomes. When it comes to governing for sustainable development in the globally interconnected context, in addition to the above characteristics, good governance requires the ability to deal with complexity and uncertainty, as well as the multiplicity of interests and perspectives necessarily involved when development is deemed 'sustainable'. Some writers term this ability 'reflexivity' (e.g. Brousseau et al., 2012; Voss & Kemp, 2006).

1.4 Crisis

The 21st-century governance context is shaped by a range of issues that arguably amount to crises, challenging the legitimacy of both economic and governance institutions. A crisis rhetoric is currently widely used in reference to climate change and biodiversity loss. While media and interest organisations such as Extinction Rebellion (XR) provide a particular slant (Grossman, 2020; Watts et al., 2019), the state of the Earth system boundaries, is increasingly referenced as a crisis also by formal governance actors (Driscoll et al., 2018; IPCC, 2023). By 2023, tens of thousands of authorities had declared climate emergencies, often accompanied by a recognition of a related emergency to do with the accelerating loss of biodiversity (Bush & Doyon, 2025). Similarly, statistical evidence of the persistent and growing inequality is widely referred to as an inequality crisis (Chancel et al., 2021; Coffey et al., 2020). The Covid-19 pandemic that gripped the world in the early years of the 2020s, killing hundreds of thousands of people, is seen as a health crisis compounded by inequality (Addley, 2021; Mishra et al., 2021).

A crisis refers to deep-seated societal problems where a prevalent trusted solution, paradigm or system is rendered inadequate by sudden or gradual, internal, or external developments. Rosenthal et al. (1989 cited in Boin, 2004, p. 167) define crisis as: 'a serious threat to the basic structures or the fundamental values and norms of a social system, which under time pressure and highly uncertain circumstances necessitates making critical decisions'. A crisis therefore marks a period of discontinuity and decline or loss of legitimacy – public trust and acceptance – of authoritative institutions (Boin, 2004).

The notion of crisis can also be seen as a rethorical device engaged by stakeholders attempting to de-legitimise the present state of affairs and to induce reform (Kuipers, 2006).

For example, the climate crisis is a term used by advocacy organisations and formal stakeholders to highlight the lack of sufficiently robust response to the mounting scientific evidence of climate change and biodiversity loss (Bush & Doyon, 2025). It is employed to counter and delegitimise powerful societal drivers of climate change, such as the fossil energy industry, which leverages the current economic system to resist a shift away from fossil energy (Christophers, 2025). XR, the grassroots and advocacy movement, has in the past decade vocally drawn attention to powerful formal institutions such as governments, banks and corporations which continue to subsidise and fund the fossil energy industry despite commitments that many of them have made to carbon neutrality by 2050. While the current economic system continues to shape the everyday context, decision-making and behaviour in the private and public realms in multiple ways, the growing use of the crisis rhetoric also by formal governance institutions can be seen as symptomatic of the gradual corrosion of its legitimacy.

Multiple crises make for a challenging governance context. Urgency and legitimacy are at the forefront of the notion of crisis, but they are not principles that combine easily. Moreover, in her book titled 'Environmental Justice in a Moment of Danger' Julie Sze (2020) suggests that we are dealing with fundamental threats to the ecological basis of our existence at a particularly dangerous time. Growing inequality and the increasingly vulnerable neoliberal economic system that itself is crisis-prone are being leveraged by populist leaders to gain political power across the world in the early decades of the 21st century (Freedom House, 2022; Grant, 2021). Together with the traction afforded by social media, this has led to a cacophony of perspectives, values, and truth-claims within which governance needs to formulate goals and select approaches and refine decision-making practices (Rhodes, 2007).

It seems that governance now also operates within the context of a crisis of knowledge. This means that decision-making can no longer rely on 'following the science' as a claim to legitimacy, but it must navigate a much more complex field of plural values and knowledge claims of various degrees of legitimacy, requiring increased reflexivity and collaboration among organisations drawn widely from public and private spheres.

1.5 Content of the book

This volume provides a critical exploration of sustainable development and its implications for governance. It will do this by first, in 'Part 1: Setting the scene: governing in the age of crises', making sense of the age of crises, sustainable development, and governance. Chapter 2 employs theories of the Anthropocene and network and risk society to set out the defining characteristics of the 21st-century multi-crises context. Chapter 3 provides a critical discussion of sustainable development and its competing iterations, the debates that surround the term and the alternative approaches that address its main critiques and aim to provide tangible ideas for transformational change. Chapter 4 explores the notion of power, its implications for good governance and key theoretical approaches to understanding efforts to govern sustainable development issues from local to the global scale.

'Part 2: Understanding global governance issues and solutions' focuses on the two sustainable development issues of climate change and food security and their governance responses. Each chapter features examples of governance responses drawn internationally, with the aim of highlighting the 21st-century governance challenges that are particularly relevant to each issue domain and that have been foregrounded in Chapter 2. Chapter 5 on climate change focuses on climate change drivers and impacts, and their governance responses that illustrate the complexity of evidence-based policy making in the context of highly politicised stakes and unequal power relations. Notions of justice and the ability of different governance approaches to engage with them are explored in detail. Chapter 6 on food security focuses on the international food system, the global challenges to food security, and the pontential of the rights-based approach to food. It examines specific governance solutions, devised in the context of networked and globalised supply chains and pricing mechanisms, which aim to reconcile the need for sufficient availability and access, health and sustainability in food production. These two 'issue chapters' reference the theories elaborated in Chapters 2–4 for the purpose of illustrating how they support a detailed understanding of key challenges and principles for good governance. Due to the limitations on space, the issue chapters do not aim to provide comprehensive explanations of climate change and food security but merely to illustrate these key sustainability challenges and their governance responses to readers who may not have much prior knowledge of the environmental domain.

Finally, 'Part 3: Lessons for good governance for sustainable development' explains how the critical understanding of governance for sustainable development developed throughout the book can be applied to the evaluation of a range of distinct types of governance efforts. Chapter 7 provides the reader with an explanation of the main challenges and principles of good governance for sustainable development. It also delves into policy evaluation and helps the reader to start applying the understanding they have gained of sustainable development and its principles into examples of governance in practice. Chapter 8 concludes on good governance for sustainable development, providing two examples of its operation in practice. Part 3 is designed to support the development of skills in governance design and evaluation, inviting the reader to apply the knowledge and understanding they have gained from the book into the formulation and application of policy evaluation criteria.

References

Addley, E. (2021, February 13). 'A very dangerous epoch': Historians try to make sense of Covid. *The Guardian* https://www.theguardian.com/world/2021/feb/13/a-very-dangerous-epoch-historians-try-make-sense-covid

Aligica, P. D., & Tarko, V. (2012). Polycentricity: From Polanyi to Ostrom and Beyond. *Governance: An International Journal of Policy, Administration, and Institutions*, *25*(2), 237–262.

Beck, U. (1992). *Risk society: Towards a new modernity*. Sage.

Beck, U. (2016). *The metamorphosis of the world*. Polity.

Bevir, M. (2011). Governance as theory, practice, and dilemma. In M. Bevir (Ed.), *The Sage handbook of governance*. SAGE Publications Ltd. https://doi.org/10.4135/9781446200964.n1

Boin, A. (2004). Lessons from crises research. In B. W. Dayton (Ed.), *Managing crises in the twenty-first century* (pp. 165–194). International Studies Review.

Brousseau, E., Dedeurwaerdere, T., & Siebenhuner, B. (Eds.). (2012). *Introduction*. MIT Press.

Bush, J., & Doyon, A. (2025). Climate emergency declarations by local governments–what comes next? *Npj Climate Action, 4*(1), 44. https://doi.org/10.1038/s44168-025-00253-2

Castells, M. (2009). *The rise of the network society: The information age - Economy, society and culture*. John Wiley & Sons, Incorporated.

Chancel, L., Piketty, T., Saez, E., & Zucman, G. (2021). *World inequality report 2022*. https://wir2022.wid.world/www-site/uploads/2022/01/Summary_WorldInequalityReport2022_English.pdf

Christophers, B. (2025). *The price is wrong: Why capitalism won't save the planet*. Verso Books.

Clark, N., & Szerszynski, B. (2020). *Planetary social thought: The anthropocene challenge to the social sciences*. Polity Press.

Coffey, C., Espinoza Revollo, P., Harvey, R., Lawson, M., Parvez Butt, A., Piaget, K., Sarosi, D., & Thekkudan, J. (2020). *Time to care: Unpaid and underpaid care work and the global inequality crisis*. Oxfam.

Crutzen, P., & Stoermer, E. F. (2000). The anthropocene. *IGBP Newsletter*, *41*, 17–18.

Driscoll, D. A., Bland, L. M., Bryan, B. A., Newsome, T. M., Nicholson, E., Ritchie, E. G., & Doherty, T. S. (2018). A biodiversity-crisis hierarchy to evaluate and refine conservation indicators. *Nature Ecology & Evolution*, 2(5), 775–781. https://doi.org/10.1038/s41559-018-0504-8

Folke, C., Polasky, S., Rockström, J., Galaz, V., Westley, F., Lamont, M., Scheffer, M., Österblom, H., Carpenter, S. R., Chapin, F. S., Seto, K. C., Weber, E. U., Crona, B. I., Daily, G. C., Dasgupta, P., Gaffney, O., Gordon, L. J., Hoff, H., Levin, S. A., ... Walker, B. H. (2021). Our future in the Anthropocene biosphere. *Ambio, 50*(4), 834–869. https://doi.org/10.1007/s13280-021-01544-8

Freedom House. (2022). *The global expansion of authoritarian rule freedom in the world 2022 highlights from freedom house's annual report on political rights and civil liberties*. Freedom House.

Grant, W. (2021). *¡Populista! The rise of Latin America's 21st century strongman*. Head of Zeus Ltd.

Grossman, E. (2020). Emergency on planet earth. *Environmental Health Perspectives, 128*(5), 054001.

Gupta, J., Liverman, D., Prodani, K., Aldunce, P., Bai, X., Broadgate, W., Ciobanu, D., Gifford, L., Gordon, C., Hurlbert, M., Inoue, C. Y. A., Jacobson, L., Kanie, N., Lade, S. J., Lenton, T. M., Obura, D., Okereke, C., Otto, I. M., Pereira, L., ... Verburg, P. H. (2023). Earth system justice needed to identify and live within Earth system boundaries. *Nature Sustainability*, *6*(6). https://doi.org/10.1038/s41893-023-01064-1

Ham, C., & Hill, M. (1993). *The policy process in the modern capitalist state*. Harvester Wheatsheaf.

HM Government. (2019). *Voluntary national review of progress towards the sustainable development goals, United Kingdom of Great Britain and Northern Ireland*. Right Vision Media.

Hough, P. (2021). *Environmental security*. Taylor and Francis. https://doi.org/10.4324/9781003082590

IPCC. (2007). *Climate change 2007: Synthesis report. Contribution of working groups I, II and III to the fourth assessment report of the intergovernmental panel on climate change* (Core Writing Team, R.K. Pachauri & A. Reisinger, Eds.). IPCC. https://www.ipcc.ch/report/ar4/syr/

IPCC. (2023). *Climate change 2023: Synthesis report. Contribution of working groups I, II and III to the sixth assessment report of the intergovernmental panel on climate change* (Core Writing Team, H. Lee & J. Romero, Eds.). IPCC. https://www.ipcc.ch/report/ar6/syr/

Jackson, T. (2017). *Prosperity without growth: Foundations for the economy of tomorrow* (2nd ed.). Routledge. https://doi.org/10.4324/9781315677453

Jordan, A., Wurzel, R. K. W., & Zito, A. &. (2005). The rise of 'new' policy instruments in comparative perspective: Has governance eclipsed government? *Political Studies*, *53*(3), 477–496. https://doi.org/10.1111/j.1467-9248.2005.00540.x

Kallis, G., Paulson, S., D'Alisa, G., & Demaria, F. (2020). *The case for degrowth*. Polity Press.

Kaul, I. (2012). Rethinking public goods and global public goods. In E. Brousseau, T. Dedeurwaerdere, & B. Siebenhuner (Eds.), *Reflexive governance for global public goods* (pp. 37–54). The MIT Press.

KOF Swiss Economic Institute. (2024). *KOF globalisation index*. Retrieved July 12, 2025, from https://kof.ethz.ch/en/forecasts-and-indicators/indicators/kof-globalisation-index.html

Kooiman, F. (2003). *Governing as governance*. Sage.

Kothari, A., Demaria, F., & Acosta, A. (2014). Buen Vivir, degrowth and ecological Swaraj: Alternatives to sustainable development and the Green Economy. *Development (Society for International Development)*, *57*(3–4), 362–375. https://doi.org/10.1057/dev.2015.24

Kuipers, S. (2006). *The crisis imperative* (pp. 9–18). Amsterdam University Press. https://doi.org/10.1515/9789048503926-002
Lélé, S. M. (1991). Sustainable development: A critical review. *World Development, 19*(6), 607–621.
Mishra, V., Seyedzenouzi, G., Almohtadi, A., Chowdhury, T., Khashkhusha, A., Axiaq, A., Wong, W. Y. E., & Harky, A. (2021). Health inequalities during COVID-19 and their effects on morbidity and mortality. *Journal of Healthcare Leadership*, *13*, 19–26. https://doi.org/10.2147/JHL.S270175
Morrison, T. H., Adger, W. N., Brown, K., Lemos, M. C., Huitema, D., Phelps, J., Evans, L., Cohen, P., Song, A. M., Turner, R., Quinn, T., & Hughes, T. P. (2019). The black box of power in polycentric environmental governance. *Global Environmental Change*, *57*, 101934. https://doi.org/10.1016/j.gloenvcha.2019.101934
Ostrom, E., Burger, J., Field, C. B., Norgaard, R. B., & Policansky, D. (1999). Revisiting the commons: Local lessons, global challenges. *Science*, *284*(5412), 278–282. https://doi.org/10.1126/science.284.5412.278
Oxfam. (2023). *Climate equality: A planet for the 99%*. Oxfam GB. https://doi.org/10.21201/2023.000001
Pierre, J., & Peters, G. (2021). *Advanced introduction to governance*. Edward Elgar Publishing.
Piketty, T. (2014). *Capital in the twenty-first century*. Harvard University Press.
Polanyi, K. (1944). *The great transformation: The political and economic origins of our time*. Penguin.
Potrafke, N. (2015). The evidence on globalisation. *The World Economy*, *38*(3), 509–552. https://doi.org/10.1111/twec.12174
Raworth, K. (2017). *Doughnut economics: Seven ways to think like a 21st century economist* (1st ed.). Random House Business Books.
Redclift, M. (2005). Sustainable development (1987–2005): An oxymoron comes of age. *Sustainable Development, 13*, 212–217. https://doi.org/10.1002/sd.281
Rhodes, R. A. W. (2007). Understanding governance: Ten years on. *Organization Studies*, *28*(8), 1243–1264. https://doi.org/10.1177/0170840607076586
Richardson, R., Steffen, S., Lucht, L., Bendtsen, B., Cornell, C., Donges, D., Drüke Drüke, Fetzer, F., Bala, B., von Bloh, v. B., Feulner, F., Fiedler, F., Gerten, G., Gleeson, G., Hofmann, H., Huiskamp, H., Kummu, K., Mohan, M., Nogués-Bravo, D., ... Rockström, J. (2023). Earth beyond six of nine planetary boundaries. *Science Advances*, *9*(37), eadh2458. https://doi.org/10.1126/sciadv.adh2458
Robra, B., & Heikkurinen, P. (2021). Degrowth and the sustainable development goals. In W. Filho, A. Azul, L. Brandelli, SalviaA. , & T. Wall (Eds.), *Decent work and economic growth* (pp. 253–262). Springer.
Rosenthal, U. E., Charles, M. T., & Hart, P. T. (1989). *Coping with crises: The management of disasters, riots and terrorism*. Charles C Thomas Publisher.
Ruggerio, C. A. (2021). Sustainability and sustainable development: A review of principles and definitions. *Science of the Total Environment*, 786. https://doi.org/10.1016/j.scitotenv.2021.147481
San Martín, W., & Wood, N. (2021). Pluralising planetary justice beyond the North-South divide: Recentring procedural, epistemic, and recognition-based justice in earth-systems governance. *Environmental Science & Policy*, *128*, 256. https://doi.org/10.1016/j.envsci.2021.12.002
Sell, S. K. (2018). Who governs the globe. In T. Weiss & R. Wilkinson (Eds.), *International organization and global governance* (pp. 77–90). Taylor & Francis Group.
Steffen, W., Richardson, K., Johan Rockström, Cornell, S. E., Fetzer, I., Bennett, E. M., Biggs, R., Carpenter, S. R., Wim, D. V., Wit, C. A. D., Folke, C., Gerten, D.,

Heinke, J., Mace, G. M., Persson, L. M., Ramanathan, V., Reyers, B., & Sverker Sörlin. (2015). Planetary boundaries: Guiding human development on a changing planet. *Science*, *347*(6223), 1259855. https://doi.org/10.1126/science.1259855

Sørensen, E. (2002). Democratic theory and network governance. *Administrative Theory & Praxis*, *24*(4), 693–720. https://doi.org/10.1080/10841806.2002.11029383

Sze, J. (2020). *Environmental justice in a moment of danger*. University of California Press.

UN General Assembly (2015) Transforming Our World: the 2030 Agenda for Sustainable Development. A/RES/70/1 United Nations. Avaiable online at: https://digitallibrary.un.org/record/3923923?v=pdf

Voss, J. P., & Kemp, R. (2006). Sustainability and reflexive governance: Introduction. In J. P. Voss, D. Bauknecht, & R. Kemp (Eds.), *Reflexive governance for sustainable development* (pp. 3–28). Edward Elgar Publishing. https://doi.org/10.4337/9781847200266.00009

Voulvoulis, N., Giakoumis, T., Hunt, C., Kioupi, V., Petrou, N., Souliotis, I., Vaghela, C., & Wan Rosely, W. (2022). Systems thinking as a paradigm shift for sustainability transformation. *Global Environmental Change*, *75*, 102544. https://doi.org/10.1016/j.gloenvcha.2022.102544

Watts, J., Blight, G., McMullan, A., & Gutierrez, P. (2019, October 9). Half a century of dither and denial – a climate crisis timeline. *The Guardian*. https://www.theguardian.com/environment/ng-interactive/2019/oct/09/half-century-dither-denial-climate-crisis-timeline?CMP=Share_iOSApp_Other

WCED. (1987). *Our common future: Report of the UN World Commission on Environment and Development*. Oxford University Press.

WEF. (2025). *The global risks report 2025 20th edition*. World Economic Forum. https://reports.weforum.org/docs/WEF_Global_Risks_Report_2025.pdf

Weiss, L. (2003). Introduction: Bringing domestic institutions back in. In L. Weiss (Ed.), *States in the global economy: Bringing the domestic institutions back in*. Cambridge University Press.

Weiss, T. G., & Wilkinson, R. (2018). From international organization to global governance. In T. Weiss & R. Wilkinson (Eds.), *International organization and global governance* (2nd ed.). Routledge, Taylor & Francis Group.

Part 1

Setting the scene – governing in the age of crises

2 The 21st-century governance context

Learning points

- What are the three dimensions of the crisis characterising the 21st-century governance context?
- What are the key drivers and manifestations of the planetary crisis and how are Earth system boundaries relevant to it?
- Why is inequality growing and how can the network society theory help us understand the unequal distribution of wealth and opportunity across the globe?
- What does the crisis of knowledge refer to and how does it relate to populism and the colonial legacy?
- What requirements do the multiple crises set for governance in the 21st century?

2.1 Introduction

As discussed in Chapter 1, 'crisis' refers to a situation where a prevalent trusted solution (for example, a social system, technology, or a scientific paradigm) is rendered inadequate by sudden or gradual developments. A crisis therefore signals the loss of legitimacy of prevalent norms and institutions (Boin, 2004). On one hand, the 21st-century crises evoke the need for sustainable development and, on the other, render attempts to achieve it more difficult. They set specific challenges for governance and call for careful consideration of the principles and practices of good governance in the 21st century.

The planetary crisis refers to the relationship between humans and the environment. In specific, it stems from the extent to which humans are currently exploiting the resources and disrupting the functions of the planetary system that constitutes the Earth, our home. This is captured in the notion of the Anthropocene, the name proposed for a new geological epoch that the Earth is said to be entering (Crutzen & Stoermer, 2000). It marks a shift to an era of instability, prone to unprecedented changes in weather, ocean currents and land systems such as rainforests like the Amazon (Clark & Szerszynski, 2020; Folke et al., 2021). This articulation of the interdependence of humans

DOI: 10.4324/9781003179207-3

and the Earth underpins the new field of Earth system science and the notion of nine Earth system boundaries (Table 2.1 in section 2.2), which are central to the stability of the Earth system (Steffen et al., 2015). One key indicator of the shift into the Anthropocene is that we have at the time of writing crossed six of the nine system boundaries and face a range of heightened environmental risks that are also destabilising our social systems (Richardson et al., 2023). One such example is climate change and associated temperature rise, sea level rise, droughts and storms, which risk some areas of the globe becoming uninhabitable for human population (IPCC, 2023).

The socio-economic dimension of the crises relates to the precarity of the global economic system and growing inequality (Piketty, 2014). There is an emerging consensus that, despite efforts to the opposite, the global neoliberal market economy has increased socio-economic inequality across the globe (Piketty, 2014; Sanders, 2021; WEF, 2025). In the context of the current economic system, unregulated market forces produce 'winners' and 'losers', visible in the concentration of income and wealth in fewer hands and a growing gap between the very rich and the poor (Chancel et al., 2026). Figure 2.1 presents the share of global aggregate income and wealth and income growth across different income groups, showing that 50% of the global population own only 2% of the aggregate global wealth and earn only 8% of the

Table 2.1 The nine Earth system boundaries

Boundary	*Manifestation*
1. Biosphere integrity	Loss of genetic and functional diversity in the biosphere. See Box 2.1 for an example.
2. Climate change	Human-induced breakdown of current planetary functions that control temperature and precipitation. See Chapter 5.
3. Stratospheric ozone depletion	Loss of the atmospheric layer protecting all life from the sun's harmful ultraviolet rays.
4. Ocean acidification	Decreasing marine biodiversity and the ability of the seas to regulate the climate and to provide nutrition.
5. Biochemical flows	Displacement of key nutrients such as Phosphates and Nitrates due to overuse in agriculture and industry. See Box 3.1 in Chapter 3 for more detail.
6. Freshwater provision	Decrease of freshwater reserves due to climate change and overuse.
7. Land system change	Changes to ecosystems and land cover, such as deforestation due to urbanisation or agriculture.
8. Atmospheric aerosol loading	Increase of harmful particles and chemicals in the air from the burning of different types of fuels and industrial and agricultural emissions.
9. Novel entities	The huge range of chemical compounds that humans have developed, some of which now permeate ecosystems and our food chains compromising human and ecosystem health.

Source: (Steffen et al., 2015, and Richardson et al., 2023).

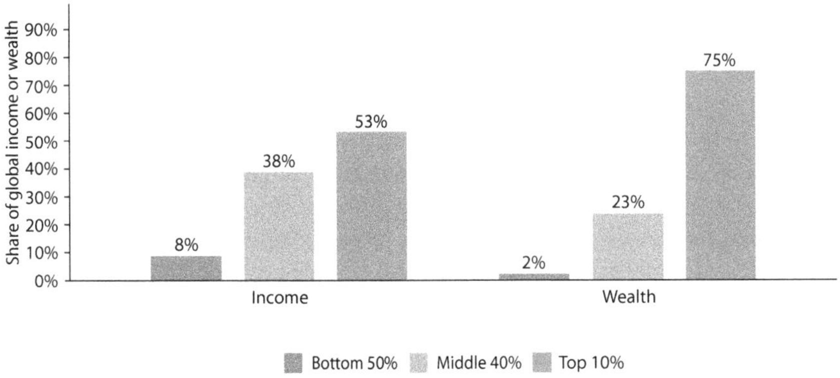

Figure 2.1 Share of global aggregate income and wealth across three income groups: the bottom 50%, the middle 40%, and the top 10%.

Adapted from Chancel et al. 2026 CC BY-NC-SA 4.0. Data from: wir2026.wid.world/methodology

aggregate income. While the amount of wealth (the sum of financial assets such as equity or bonds, and non-financial assets, such as housing or land owned by individuals) has been growing for all income groups over the three decades from 1995, Chancel et al. (2026) also show that growth has been by far the fastest for the richest 0.001%.

Inequality is also visible at the global scale, as countries display differing ability to benefit from the free-market economy. Some act as global agenda setters leading the development of new economic opportunities, while others remain sites of extraction of resources and cheap labour (Chancel et al., 2026). These divisions tend to reflect the power relations established in the colonial era.

The third dimension of the crisis refers to the politicisation of knowledge. Social media has facilitated deliberate, targeted, and instrumental use of misinformation to manipulate political allegiance, including alleged foreign intervention in elections (Enock et al., 2024). The rise of populist politicians like Mr Boris Johnson and President Donald Trump in the UK and US respectively has led to accusations of blatant use of misinformation by mainstream politicians and the need for regular 'fact checking'[1] of political speeches and announcements by the press. Beyond the cynical manipulation of knowledge towards political ends, there is another more legitimate challenge to the hegemony of 'expert' knowledge today. With evolving cultural beliefs and plural voices amplified by accessible information technology, society has become more diverse and complex in terms of identities and values. At the global scale, so-called 'epistemic colonialism' refers to the dominant position and in-built bias of scientific knowledge, which is based on a predominantly Western perspective and shaped by exclusive validity criteria (San Martín & Wood, 2021). It is increasingly recognised that a broader understanding of valid knowledge is now required, with the inclusion of Indigenous, experiential, and practical knowledge for legitimate governance (San Martín & Wood, 2021).

In the 21st century, governance for sustainable development needs to respond to the urgency implied by the planetary crisis while addressing growing inequality in a context of highly politicised knowledge and values. As set out in Chapter 1, governance is defined as *the institutions and practices of decision making and intervention regarding issues of public interest, that is undertaken by formal and informal governance actors, often situated in network relations of power rather than in a hierarchical top-down bureaucracy* (Morrison et al., 2019; Sørensen, 2002). This chapter examines the three dimensions of the crises context and explores their implications for governance in more detail.

2.2 The Anthropocene, risk society, and the planetary crisis

Ulrich Beck captures many of the central features of the present-day crises, particularly the planetary crisis, in his theory of the risk society (Beck, 1992). Beck suggests that our present predicament is characterised by the proliferation of risks – issues threatening human wellbeing and the stability of our societies. He suggests that many of these risks are rooted in industrialisation and the model of development as 'modernisation' dating to the 19th and 20th centuries (Clark & Szerszynski, 2020). While many people are now reaping the rewards of this development in terms of access to modern medicine, plentiful food and energy supply, and convenient lifestyles, this has come at a price of new risks (Beck, 1996). Industrialisation and modernisation were leveraged by ever-increasing natural resource appropriation and subjugation of regions, countries, and people in extractive and unequal relationships of appropriation and exchange, many of which remain and continue to shape the risk society today (Ghosh, 2021; Horner, 2020; Piketty, 2014). This underpins a society facing risks associated with both overabundance and scarcity.

Extensive resource extraction and transformation into energy and goods and the resulting abundance of waste and polluting emissions are now impacting every aspect of our environment. The climate crisis – resulting from excessive emissions of so-called greenhouse gases (GHGs) into the atmosphere – is evidenced by a multitude of scientific research. The Intergovernmental Panel on Climate Change (IPCC), comprising leading scientists from across the world, provides regular Assessment Reports evidencing and forecasting the impact of rising global temperature (IPCC, 2007, 2014). The latest such report published in 2022 projects intensifying precipitation and flooding interspersed with severe droughts in Europe, Africa, Australasia and North, Central and South America, and intensifying cyclones and storms as well as fire weather in the tropics and arid areas (IPCC, 2022). Despite committing to limiting global warming to 1.5°C in 2015 (UNFCCC, 2016), governments have failed to instil sufficient cuts to emissions, and the Earth passed the 1.5°C limit in 2024 (King & Cassidy, 2025).

The state of biodiversity and ecosystems has been in decline for decades (Dasgupta, 2021; Driscoll, Don A. et al., 2018). A 2019 assessment by the Intergovernmental Science-Policy Platform on Ecosystem Services (IPBES) found that 14 of 18 categories of nature's essential functions have deteriorated since 1970 (IPBES, 2019). The research on Earth system boundaries (e.g. Steffen et al., 2015; Folke et al., 2021), reveals the key pressures: land system change, climate change, and an abundance of 'novel entities' (human-manufactured chemicals). Box 2.1 explains how the loss of biodiversity poses a risk to food security as well as the ability of nature to respond to external pressures such as climate change.

Box 2.1: Loss of biodiversity as a collective risk to food security

The way most people living in the so-called developed country context produce and source food is currently profoundly industrial and global (Marsden et al., 2019). Modern agriculture is driven by the need to extract the largest possible quantity of produce from a limited land area while responding to the nutritional and preference needs of the rapidly growing global population. In most farming landscapes, this has led to the replacement of natural ecosystems with vast monocultures of provisioning species, food crops such as wheat, rice, oil palm or cattle intended for the sole purpose of food production (Dasgupta, 2021).

While this enables us to do away with one risk – food scarcity – it replaces the natural biodiversity of ecosystems with monocultures of food crops and is based on management practices that impoverish soils of ingredients that are not essential to the mass production of the food crop at hand (FAO et al., 2020). This is not just a direct risk to the diversity of species, but a development that threatens the resilience of food production itself and renders us vulnerable to sudden shocks that may lead to food insecurity (IPBES, 2019).

Genetic diversity is crucial for the resilience – ability to withstand adverse events and to maintain functionality – of a species or an ecosystem or indeed a society. But productivist agriculture focused on extracting maximum productivity with minimum inputs engages with a narrow range of genotypes (specimens of a species such as the wheat plant that have very similar genes and therefore similar characteristics) that display high productivity. This exclusivity discards the natural diversity in the genetic range, traits, community composition, population and functional and structural capacity that is naturally present within a single species (Dasgupta, 2021). This narrowing of characteristics and variation within a species renders it vulnerable, potentially unable to cope with sudden threats such as the spread of previously foreign pests or diseases, perhaps emerging because of changes in climatic conditions.

Therefore, the biodiversity loss resulting from intensive food production does not just refer to the loss of landscape beauty or iconic species such as orangutans but constitutes a much more severe risk to the long-term ability of soil and the prevalent crops to continue producing in the face of unexpected stressors. This is a collective risk to food security globally (Beck, 1996).

The Earth system boundaries displayed in Table 2.1 can be thought of as the nine most significant environmental issues that we are faced with and that are all central to our planet's equilibrium and to our survival on it (Richardson et al., 2023; Steffen et al., 2015).We have now crossed six of the nine Earth system boundaries and, as a result, the planetary system has become unstable (Richardson et al., 2023; Steffen et al., 2015). We've entered a state of deep uncertainty regarding the impacts of this instability. The notion of the Anthropocene articulates human existence as subject to the Earth system and underlines the role of humans in driving its growing instability, marking an epoch of growing socio-ecological risks (Crutzen & Stoermer, 2000). In the quest for 'development', humans have tampered with nature to produce complex risks that we not only struggle to address, but even to properly define or understand (Beck, 2016).

The theory of Risk Society provides some 'tools' to begin to make sense of the societal implications of the planetary crisis and to embrace it as a prevalent social and political phenomenon in the Anthropocene. While many of the planetary risks are collective pertaining to the whole human population (see Box 2.1), the notion of unequal risk positions (Beck, 1992) refers to how some individuals, groups or countries are better able to buffer themselves against these risks or to escape them altogether thanks to their ability to build resilience and adapt (Oxfam, 2023). Resilience and the capability to adapt depend on both natural and human capitals, and unequal risk positions can often be traced back to power relations established in the colonial era (Ghosh, 2021). This is evident, for example, in the case of climate change, an impact of the deposition of greenhouse gases (GHGs) stemming from human activities associated with the industrial era that started in Europe in the 1700s. The front-runners of industrialisation now form the largest economies and constitute the largest historical emitters of GHGs, but the most severe impacts of climate change in the form of destructive tropical storms and droughts that compromise infrastructure and food security are felt in developing countries (IPCC, 2007, 2014, 2018).

Beck (1992, 2006) suggests that we are now living in an era where we are increasingly confronted by the consequences of the values and actions characteristic of 'modernity' that were aimed at bringing technological development and increased wealth to benefit all. It has become evident that Western science-led technological development and top-down hierarchical decision-making regarding the goals and means of development have not led to desirable outcomes for all (Beck, 2006; Voß & Kemp, 2006). The environmental and socio-economic dimensions of the present-day crises attest to that. Therefore, we now need to reconsider the meaning of development in the light of a broader set of values, knowledges, and risk positions that characterise society today.

In terms of governance, the planetary crisis underlines some key challenges. We need science to understand the causes and consequences of crossing the Earth system boundaries, but the resulting risks are profoundly political in that they are caused by human activities and evoke unequal risk positions. They raise questions of ethics, justice and responsibility. The challenge is not only to understand how to mitigate these risks but how to do this in a just manner calling to task those that have caused them and building the resilience of those most affected by them. If left unaddressed, the deterioration of planetary functions is likely to further exacerbate existing inequalities, threatening to form a vicious cycle that further destabilises our societies (Folke et al., 2021; Raworth, 2017). The next section will consider the drivers and implications of the persistent and growing inequality that characterises the 21st century.

2.3 The network society, growth imperative, and inequality

Thanks to globalisation (Box 2.2), we live in an increasingly interconnected and therefore also interdependent world (Ritzer & Dean, 2015). What happens at a local scale matters at the global scale and vice versa. The globalised commodity networks through which food products travel from farmers to consumers provide an example of this. In 2017, a 'courgette crisis' of shortage of produce in UK supermarkets (Weaver, 2017) was caused by local weather and pricing decisions made by marketing organisations in Spain. Similarly, UK consumers of Spanish horticultural products are implicated in the degradation of coastal wetlands caused by the intensive water use by vegetable producers in Spain (Juntti & Downward, 2017).

Box 2.2: Globalisation

As defined by Ritzer and Dean (2015, p. 2), globalisation is: 'a trans planetary process or set of processes involving increasing liquidity and the growing multidirectional flows of people, objects, places and information as well as the structures they encounter and create that are barriers to, or expedite, those flows'. While people, goods, and money have for long been transported from place to place, today's 'global era' marks a difference in the ease, frequency, and speed with which this is happening. There are, however, barriers to these global flows also. These may be unintended factors 'blocking' movement, such as difficulty in accessing suitable technology due to issues of affordability or availability (this often creates inequalities between people where fast broadband or long-haul flights are not at the disposal of the poor or those living in remote rural locations, for example). But there are also intentional barriers to such flows, such as immigration regulations, and standards and taxes for imported goods. These flows and barriers are part and parcel of globalisation, which according to many has become something of an unstoppable force driven by the myriad of advantages and opportunities that it offers to the economy and society at large.

These global flows have taken on an almost self-perpetuating nature, and the interconnectivity and dependence that they engender also present risks. Ritzer and Dean (2015) refer to the difficulty of controlling the flow of digital content across national borders and the subsequent difficulty of prosecuting any perpetrators of digital crime. In 2020, we witnessed the threat that global liquidity of viruses poses in the form of a global pandemic. The ease with which the virus (SARS-CoV-2) causing the dangerous Covid-19 infection spread throughout the world from its likely origins in China to constitute a global pandemic exemplifies a tangible threat associated with globalisation.

Therefore, globalisation has a myriad of manifestations and impacts in our everyday lives, related to the transfer of people and material goods but also of ideas, organisms and practices that we now take for granted as shaping our everyday lives.

The way in which society and the global economic system are currently structured has a huge influence on how income, goods and environmental problems such as water scarcity or pollution are distributed across the globe. Two theories are key to understanding how inequality is produced in the global socio-economic system: the theory of network society and the capitalist neoliberal global market economy. 'Network society' (Castells, 2004, 2009) refers to the networked and interconnected structure of today's society. It offers means to understand how agency and power are mediated by technology and orchestrated by economic forces in the global economic and supply systems.

In the capitalist market economy (Piketty, 2014), resources are manufactured into useful outputs (goods) to fulfil human needs by engaging with 'capital' (e.g. infrastructure and technology), 'labour' (provided by humans), and natural resources, as depicted in Figure 2.2. In a neoliberal, relatively unregulated market, these manufactured private goods are sold at a price that reconciles demand (utility value) and cost of production. This produces two classes of people: the capital-owning class and labourers.

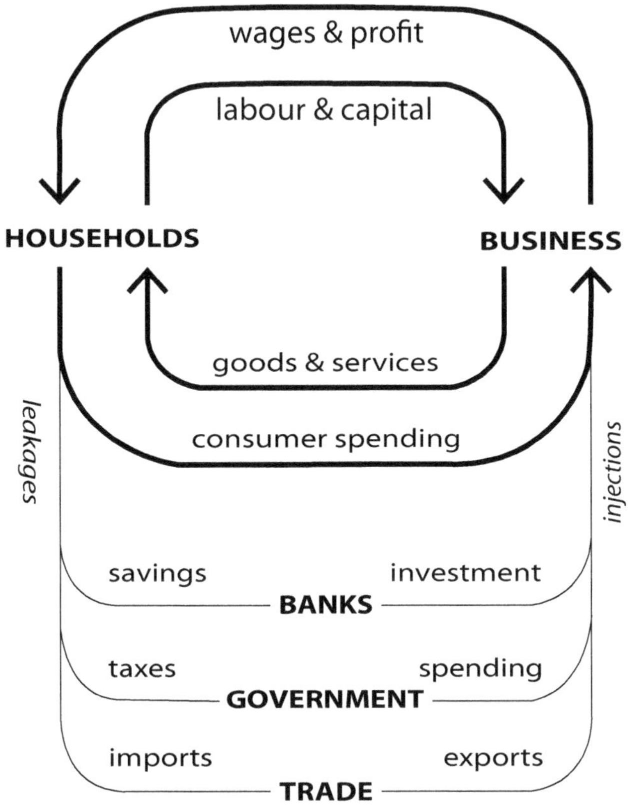

Figure 2.2 The circular flow diagram depicting the economic system where capital and resources are combined to produce goods and services in exchange for labour.

Adapted from Raworth 2017. Copyright © 2017 Kate Raworth.

Thomas Piketty (2014) explains how the way that income is produced in the current neoliberal capitalist global market economy leads to the accumulation of profit in the hands of the people who own the capital means of production. Therefore, if left unregulated, this economic system is inherently unequal, with wages failing to match the growth in profit from capital and rent. Taxation and public expenditure on education and health care, can fucniton as effective redistribution mechanims to combat income-related inequality, but they are very unevenly used across the world, and the richest 10% evade them, paying proportionately much less tax than the rest of income groups (Chancel et al., 2026). This largely explains the accumulation of 75% of global wealth among the top 10% of people in Figure 2.1.

The international food system serves as an example of how the capitalist market economy operates to produce inequality in the context of the network society. The food system refers to the global network through which food travels to reach consumers (Barrett et al., 2004). Everyday food items that we consume, for example coffee, mask a multiplicity of contributors behind an attention-grabbing logo of one corporate brand. Coffee beans travel from producers (farmers) through wholesalers, manufacturers, and retail outlets to consumers, and pass through the hands of logistics companies, import and export agencies and various regulators along the way. Due to the availability of technology and highly skilled workers, the corporation behind the brand is usually based in a developed country such as Switzerland or the US. It hosts the most high-value stages of the commodity network (in this case roasting, grinding and branding of coffee).

Large supermarkets also based in developed countries are in a powerful position in terms of price setting, whereas the often small-scale coffee bean producers in middle- or low-income countries such as Vietnam or Colombia have to comply with these internationally set prices and product standards that meet import regulations, corporate criteria and consumer preferences in the destination countries (Reinecke et al., 2012). In this way, commodity networks shape and maintain an international division of labour, which replicates the power imbalances dating back to the colonial era, where the countries and labourers with advanced technology and highly skilled labour reap the majority of the profits. In the meantime, countries depending on the exports of primary goods such as agricultural products remain in the low to middle-income category (Castells, 2009). Box 2.3 exemplifies the implications of this kind of marginalisation for small-scale farmers.

Box 2.3: Balancing the need to address socio-economic marginalisation and environmental sustainability with the need for economic transformation to gain global market share in rural India

Hemalata Dandekar (2019) describes how, following independence from the UK in 1947, India has continued to implement a market-led approach to rural development. But this strategy has failed to address the marginalisation of its vast and predominantly poor subsistence farming communities. Rural planning efforts have centred on catering to the plural needs of the diverse class (caste) based rural communities requiring attention to equity in access to infrastructure, education, housing and amenities for all, including the landless and very poor agricultural workers and subsistence farmers. But these ambitions have been compromised by the need to ramp up agricultural competitiveness in the global arena. Early on, the need for foreign exchange led to policy-driven structural transformation towards bigger farms equipped with irrigation and cultivation technologies and chemical inputs striving to produce the most resilient and productive crop varieties. This export-led growth model (Gray, 2014) underpinned corporatisation of farms and exclusion of those that could not adopt this 'modern' model of production. Many rural poor subsequently fled to cities, creating informal 'slum' economies. While agriculture did become more productive and commodification opened opportunities for more diverse income streams, this was informed by price and demand signals from global markets, rather than by the needs of rural communities, which became even more polarised in terms of socio-economic assets, incomes and opportunities.

The aspiration to maintain and grow the global market share of Indian agricultural products continues to trump the needs of the rural poor, failing social justice aspects such as access to public services and environmental sustainability (Dandekar, 2019). In the era of the climate crisis, this is not only unfair but dangerous, as rural communities are having to cope with growing challenges to land and crop productivity from climate change impacts. Resilience, the ability to cope with sudden shocks as well as incremental change in weather patterns, depends on access to diverse assets and socio-economic safety nets such as unemployment benefits and emergency aid. Poor landless communities continue to lack such entitlements, which renders the Indian countryside highly vulnerable. Moreover, export-led production often overlooks impacts of intensive production on the local resource base. This is visible in water shortages and a decrease in water quality in many regions due to over-extraction for intensive mechanised agriculture, leaving rural communities often water poor.

The network society theory draws attention to the way that digitalisation often exacerbates these power imbalances. In the 21st century, the capitalist relations of investment, production and exchange are increasingly facilitated by digitalisation (Morgan et al., 2008), which accelerates the production and travel of tangible goods through commodity networks and contributes to efficiency and productivity. Digitised centralised trading has enabled a truly global commodity market, where technology facilitates investment decision-making with computers reading signals regarding asset value and profitability and making decisions aimed at maximising profit. So-called 'derivatives' or 'futures' are traded by hedge funds, and consist of commodities such as oil, wheat, coffee or rice. Commodity values are influenced by speculative behaviour rather than actual availability or supply of the commodity in question. These digital networks (Castells, 2009) therefore further amplify the power relations. Those who have the skills and access to digital technology and speculative stock markets can leverage this technology and shift agency away from producers to those merely seeking to make profit and thereby becoming the main agents for value creation.

The network metaphor can also be applied to understand how corporations and employment conditions are evolving in the 21st century to uncover further drivers of inequality. Castells (2004, 2009) argues that in search of efficiency gains to ensure competitiveness and viability in accelerating markets, corporations themselves have become networked. They no longer consist of large stable structures that house all stages of the production process in one company, but increasingly display small core executives that rely on satellites – external providers of content, goods and services.

For example, smartphones are not produced by one company but just like coffee, the design and branding are done by a core company while parts and software are purchased from smaller suppliers in various different predicaments. These contributions are valued not just according to consumer willingness to pay but the demands of competitiveness and profit extraction, leading to a precarious small-business environment. The so-called gig or platform economy is another iteration of the networked corporation. Platforms such as Uber or Deliveroo bypass a central corporate structure altogether and provide a flexible and low-cost business model in the context of fluctuating demand for services, while offering minimal job security and worker protections (Graham et al., 2020).

While the network society produces a plurality of opportunities for innovation, flexibility and efficiency, driven by the neoliberal capitalist market logic, it also produces a plethora of risks that underpin inequality (Castells, 2004). Digitised commodity networks and networked business models guided by competitiveness and profit maximisation have produced individuals and countries that face heightened risks of marginalisation and environmental harms. This is visible, for example, in the regional disparity in monthly income depicted in Figure 2.3. But Chancel et al. (2026) emphasise that the wealth disparity between countries and regions is not the result of

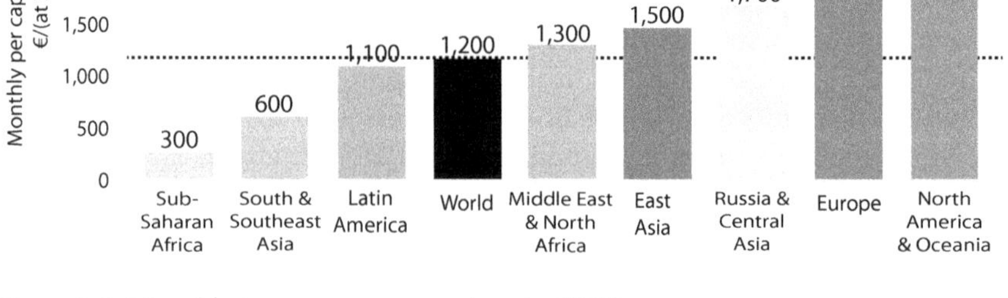

Figure 2.3 Monthly incomes across regions in 2025.

Adapted from Chancel et al. 2026. CC BY-NC-SA 4.0. Data from wir2026.wid.world/methodology.

free market forces alone but is reinforced by policies, such as the powerful position of the US Dollar as a reserve currency and decisions by international credit rating agencies. These governance instruments systematically favour wealthy economies such as the US, EU and UK.

In terms of governance, a key challenge is capturing these drivers of inequality and planetary instability vested in complex systems and devising legitimate interventions that deliver equitable and fair outcomes. In addition to the core principles of accountability, transparency, plural representation and deliberation, systems thinking and reflexivity, embracing complexity and remaining flexible in the face of unexpected outcomes, emerge as key criteria for good governance (Brousseau et al., 2012; Voulvoulis et al., 2022).

2.4 Complexity, uncertainty, and the crises of knowledge

The planetary and inequality crises both result in complex risks that require access to both expert knowledge and observation and experience from the context of a plural range of everyday lives and shared values (Beck, 1992). This means that there are no obvious 'correct solutions' to key societal problems such as climate change or food security, and that any solution will evoke conflicts of values, perspectives and interests. Such 'wicked' issues tend to be highly politicised and place difficult demands for governance interventions, particularly in regard to the evidence base that decision-making engages with (Turnpenny et al., 2009).

For example, the risk that biodiversity depletion causes to food security (Box 2.1) represents such a typical 'wicked challenge' requiring both expert understanding of complex environmental phenomena and experiential knowledge reflecting the diverse contexts within which farmers operate. Attention to biodiversity will likely compromise the economic interests of a range of actors such as large-scale intensive farmers who focus on maximising crop yields in line with conventional good farming practice, and manufacturers of pesticides whose products may encounter declining demand. While the scientific assessment of biodiversity loss indicates that it is a pressing issue that requires immediate action (FAO et al., 2020), in practice this 'scientific way of knowing' will be juxtaposed with the business perspective and 'practical knowledge' that frames the risk of biodiversity loss as 'relative' to economic risk at farm level (Beck, 1992). This plural nature of knowledge, where no one way of knowing is wrong, but all of them illuminate different perspectives on any given risk or challenge, is both tricky and necessary to deal with in governance.

The inequality crisis underpins further politicisation of both knowledge and governance interventions. This is associated with rising discontent. Since the late 20th century, the faith placed by decision-makers in free markets for both goods and finance has compromised some income and wealth redistribution mechanisms and public services such as social security and public health care (Alberti et al., 2018). According to Castells (2004) this includes also the shift towards the networked model of corproation and attendant loss of the 'paternalistic' employer offering a job for life and a pension. The economic crises of the 1990s, 2000s, and 2020s, and the subsequent austerity politics, only accelerated the resulting precarity, leading to inflation and interest rate rises triggering a 'cost of living crisis' (Economist Intelligence Unit, 2024). In this context, social justice implications of governance inteventions need to be carefully managed. For example, the 2018 'Yellow vest protests' in France were triggered by a carbon tax on petrol, intended to incentivise the transport industry to cut GHG emissions (Driscoll, 2023). But in the context of already precarious working conditions, it was experienced as grossly unfair and sparked country-wide demonstrations.

This kind of discontent makes for a situation rife for exploitation by those who are able to capture the attention of people in search of strong leaders in difficult times (Grant, 2021). In politics, we are seeing this in the rise of so-called populism, where opportunistic leaders surge to power on the back of unfounded promises speaking to people's fears and insecurities. Populist leaders, such as the UK's Boris Johnson and Jair Bolsonaro of Brazil, often sow further mistrust of knowledge elites or 'experts' and seek to suppress the voices of those who question their aims or means whether in the press or the Parliament (Grant, 2021). This is associated with a significant decrease in democratic freedoms globally, and in many countries, of the central structures of society that are, despite their well-recognised limitations, crucial for enabling plural voices to be heard and upholding accountability of the political elite (Freedom House, 2022; Sze, 2020).

For example, limits to press freedom and access to information, as well as the dismantling of independent judiciaries and freedom of speech typical of authoritarian regimes, are increasingly seen in formerly strong democracies like the USA and the UK (Freedom House, 2022). Together with social media affording unlimited user-created content and developments in Artificial Intelligence, this amounts to 'a period of discontinuity' marking a distinct decline in public trust and acceptance of the values and institutions of knowledge production, processing, sharing, and appropriation (Boin, 2004). Governments struggle to find a balance between safeguarding freedom of speech and curbing so-called hate speech that incites discrimination and violence. We are in an epistemological crisis, where previously valued knowledge institutions such as experts and the media are no longer able to deliver on public expectations due to changes in these internal and external factors.

Selective use of information has of course always been central to politics, as well as advertising, and politicians such as the UK Prime Minister Boris Johnson and the US President Donald Trump discussed above have employed these tactics throughout the ages. But the complex new and mounting uncertainties of the 21st century seem to have made society even more vulnerable to this type of politicisation of knowledge.

Amitav Ghosh (2021) highlights a further dimension of the crisis of knowledge and like Beck (2006), reminds us that what the currently dominant form of Western scientific knowledge is by no means value neutral. With industrialisation, European nations began the systematic conquest and subordination of nature to human needs. Western science and institutions of knowledge production were established and funded to serve this quest. Science continues to shape our understanding of the impact of this – the complex cause-consequence chains that were set in motion by this appropriation of nature and extraction of natural resources at scale – and provides an important but limited perspective incapable of recognising many of the losses experienced along the way (Clark & Szerszynski, 2020). In order to alleviate the planetary crisis and the crisis of inequality, ideas, values, and tacit norms and biases vested in the 'modern truths' that dominated since the industrial era need to be critically interrogated. Some of them will need to be adjusted by integrating perspectives from more representative epistemologies, for example, those represented by Indigenous knowledge (Beck, 2006; Ghosh, 2021; Sze, 2020). But we must be careful not to confuse this plurality that is sensitive to difference with populism, the cynical use of information to agitate segments of communities for political gain.

2.5 Conclusions

This chapter focused on the notion of multiple crises as a defining feature of the 21st-century governance context. A crisis refers to a state where the structural (institutional and rules-based) basis of society has proven insufficient to guarantee functionality and is suffering a loss of legitimacy. Three crisis dimensions can be identified as pertaining to the planetary, socio-economic, and knowledge-related developments since the latter half of the 20th century. The world is now grappling with multiple challenges that underline the need for sustainable development and set a range of requirements for governance.

The notion of risk now permeates every aspect of our societies, in the form of unprecedented environmental events, an increasing number of people in relative or abject poverty, and increasing discontent and manipulative use of disinformation by domestic and foreign actors to influence political power. Two key drivers of the risks emerge: the prevalent neoliberal and growth-oriented economic system and the increasingly fragmented networked model of society and business that disrupts the stability of values, institutions, and livelihoods. In the context of the growing instability of the planetary system and associated complex risks such as climate change and food insecurity, the scope for disinformation has grown. Simultaneously, 'solutions' such as export-led growth and the gig economy, which leave communities and individuals vulnerable to exploitation, have led to the marginalisation of vast groups of people who feel left out of society. This has rendered people vulnerable to populist leaders who cynically exploit this discontent to gain power but often fail to deliver on community needs as promised.

All this poses a challenge to the legitimacy of governance interventions for sustainable development, as *fundamental societal transformation towards equitable and regenerative relations among people and the planet, addressing the core drivers of unsustainability.* As society is becoming more diverse and globally interconnected, there is a need to integrate systems thinking and broader representation into decision-making on issues of common interest. The scientific perspective remains crucial for understanding and addressing the prevalent crises, but there is a need to recognise the validity of different forms of knowledge beyond the Western scientific perspective that stems from colonial power relations and practices. Subsequent chapters will delve into how to define sustainable development in this challenging era and explore theories, approaches and case studies that help us decipher the meaning of good governance for sustainable development in these fraught circumstances.

Note

1 See for example https://www.factcheck.org/2020/11/checking-the-facts-in-the-world-of-trump/

References

Alberti, G., Bessa, I., Hardy, K., Trappmann, V., & Umney, C. (2018). In, against and beyond precarity: Work in insecure times. *Work, Employment and Society*, *32*(3), 447–457. https://doi.org/10.1177/0950017018762088

Barrett, H. R., Browne, A. W., & Ilbery, B. W. (2004). From farm to supermarket: The trade in fresh horticultural produce from sub-Saharan Africa to the United Kingdom. In HughesA. & ReimerA. (Eds.), *Geographies of commodity chains*. Routledge.

Beck, U. (1992). *Risk society: Towards a new modernity*. Sage.

Beck, U. (1996). World risk society as cosmopolitan society: Ecological questions in a framework of manufactured uncertainties. *Theory Culture & Society*, *13*(4), 1–32.

Beck, U. (2006). Reflexive governance: Politics in the global risk society. In *Reflexive governance for sustainable development*. Edward Elgar Publishing. https://doi.org/10.4337/9781847200266.00011

Beck, U. (2016). *The metamorphosis of the world*. Polity.

Boin, A. (2004). Lessons from crises research. In B. W. Dayton (Ed.), *Managing crises in the twenty-first century* (pp. 165–194). International Studies Review.

Brousseau, E., Godard, O., Dedeurwaerdere, T., Siebenhuner, B., Kaul, I., Arce, D. G., Frey, B. S., Gunningham, N., Sarr, M., & Swanson, T. M. (2012). *Reflexive governance for global public goods*. MIT Press.

Castells, M. (2004). *The network society: A cross-cultural perspective*. Edward Elgar Publishing Limited.

Castells, M. (2009). *The rise of the network society: The information age - economy, society and culture*. John Wiley & Sons, Incorporated.

Chancel, L., Gómez-Carrera, R., Moshrif, R., & Piketty, T. (2026). *World inequality report 2026*. https://wid.world/document/world-inequality-report-2026/

Clark, N., & Szerszynski, B. (2020). *Planetary social thought: The anthropocene challenge to the social sciences*. Polity Press.

Crutzen, P., & Stoermer, E. F. (2000). The anthropocene. *IGBP Newsletter*, *41*, 17–18.

Dandekar, H. C. (2019). Rural planning in an emerging economy: India. In ScottM., N. Gallent, & M. Gkartzios (Eds.), *The Routledge companion to rural planning*. Routledge.

Dasgupta, P. (2021). *The economics of biodiversity: The Dasgupta review abridged version*. HM Treasury.
Driscoll, D. A. (2023). Populism and carbon tax justice: The yellow vest movement in France. *Social Problems*, *70*(1), 143–163. https://doi.org/10.1093/socpro/spab036
Driscoll, D. A., Bland, L. M., Bryan, B. A., Newsome, T. M., Nicholson, E., Ritchie, E. G., & Doherty, T. S. (2018). A biodiversity-crisis hierarchy to evaluate and refine conservation indicators. *Nature Ecology & Evolution*, *2*(5), 775–781. https://doi.org/10.1038/s41559-018-0504-8
Economist Intelligence Unit. (2024). Worldwide cost of living: How soaring inflation has affected prices globally. *The Economist*.
Enock, F., Bright, J., Stevens, F., Johansson, P., & Margetts, H. (2024). *How do people protect themselves against online misinformation? Attitudes, experiences and uptake of interventions amongst the UK adult population*. The Alan Turing Institute. https://www.turing.ac.uk/sites/default/files/2024-05/how_do_people_protect_themselves_from_misinformation.pdf
FAO, ITPS, GSBI, SCBD, & EC. (2020). *State of knowledge of soil biodiversity - Status, challenges and potentialities*. FAO.
Folke, C., Polasky, S., Rockström, J., Galaz, V., Westley, F., Lamont, M., Scheffer, M., Österblom, H., Carpenter, S. R., Chapin, F. S., Seto, K. C., Weber, E. U., Crona, B. I., Daily, G. C., Dasgupta, P., Gaffney, O., Gordon, L. J., Hoff, H., Levin, S. A., ... Walker, B. H. (2021). Our future in the Anthropocene biosphere. *Ambio*, *50*(4), 834–869. https://doi.org/10.1007/s13280-021-01544-8
Freedom House. (2022). *The global expansion of authoritarian rule freedom in the world 2022 highlights from Freedom House's annual report on political rights and civil liberties*. Freedom House.
Ghosh, A. (2021). *The Nutmeg's Curse parables for a planet in crisis*. University of Chicago Press. https://doi.org/10.7208/chicago/9780226815466
Graham, M., Woodcock, J., Heeks, R., Mungai, P., Van Belle, J., du Toit, D., Fredman, S., Osiki, A., van der Spuy, A., & Silberman, S. M. (2020). The fairwork foundation: Strategies for improving platform work in a global context. *Geoforum*, *112*, 100–103. https://doi.org/10.1016/j.geoforum.2020.01.023
Grant, W. (2021). *¡Populista! The rise of Latin Amrica's 21st century strongman*. Head of Zeus Ltd.
Gray, D. (2014). Economic approaches to the rural. In G. Bosworth & P. Somerville (Eds.), *Interpreting rurality: Multidisciplinary approaches* (pp. 32–53). Routledge.
Horner, R. (2020). Towards a new paradigm of global development? Beyond the limits of international development. *Progress in Human Geography*, *44*(3), 415–436. https://doi.org/10.1177/0309132519836158
IPBES. (2019). *Global assessment report on biodiversity and ecosystem services of the Intergovernmental Science-Policy Platform on Biodiversity and Ecosystem Services (Version 1)*. https://doi.org/10.5281/zenodo.6417333
IPCC. (2007). *Climate change 2007: Synthesis report. Contribution of working groups I, II and III to the fourth assessment report of the intergovernmental panel on climate change* (Core Writing Team, R. K. Pachauri & A. Reisinger, Eds.). IPCC. https://www.ipcc.ch/report/ar4/syr/
IPCC. (2014). *Climate change 2014: Synthesis report. Contribution of working groups I, II and III to the fifth assessment report of the intergovernmental panel on climate change* (Core Writing Team, R. K. Pachauri & L. A. Meyer, Eds.). IPCC. https://www.ipcc.ch/report/ar5/syr/
IPCC. (2018). *Global warming of 1.5°C. An IPCC special report on the impacts of global warming of 1.5°C above pre-industrial levels and related global greenhouse gas emission pathways, in the context of strengthening the global response to the threat of climate change, sustainable development, and efforts to eradicate poverty*. Intergovernmental Panel on Climate Change. https://www.ipcc.ch/sr15/

IPCC. (2022). *Climate change 2022: Impacts, adaptation, and vulnerability. contribution of working group II to the sixth assessment report of the intergovernmental panel on climate change*. Cambridge University Press. https://www.ipcc.ch/report/sixth-assessment-report-working-group-ii/

IPCC. (2023). *Climate change 2023: Synthesis report. Contribution of working groups I, II and III to the sixth assessment report of the intergovernmental panel on climate change* (Core Writing Team, H. Lee & J. Romero, Eds.). IPCC. https://www.ipcc.ch/report/ar6/syr/

Juntti, M., & Downward, S. D. (2017). Interrogating sustainable productivism: Lessons from the 'Almerían miracle'. *Land Use Policy*, *66*, 1. https://doi.org/10.1016/j.landusepol.2017.04.016

King, A., & Cassidy, L. (2025). Earth is already shooting through the 1.5°C global warming limit, two major studies show. *The Conversation*. Retrieved September 30, 2025, from http://theconversation.com/earth-is-already-shooting-through-the-1-5-c-global-warming-limit-two-major-studies-show-249133

Marsden, T., Faus, A., & Sonnino, R. (2019). Reproducing vulnerabilities in agri-food systems: Tracing the links between governance, financialization, and vulnerability in Europe post 2007–2008. *Journal of Agrarian Change*, 19, 82–100. https://doi.org/10.1111/joac.12267

Morgan, K., Marsden, T., & Murdoch, J. (2008). Networks, conventions, and regions: Theorizing 'worlds of food'. In K. Morgan, T. Marsden, & J. Murdoch (Eds.), *Worlds of food: Place, power, and provenance in the food chain* (pp. 7–25). Oxford University Press.

Morrison, T. H., Adger, W. N., Brown, K., Lemos, M. C., Huitema, D., Phelps, J., Evans, L., Cohen, P., Song, A. M., Turner, R., Quinn, T., & Hughes, T. P. (2019). The black box of power in polycentric environmental governance. *Global Environmental Change*, *57*, 101934. https://doi.org/10.1016/j.gloenvcha.2019.101934

Oxfam. (2023). *Climate equality: A planet for the 99%*. Oxfam GB. https://doi.org/10.21201/2023.000001

Piketty, T. (2014). *Capital in the twenty-first century*. Harvard University Press.

Raworth, K. (2017). *Doughnut economics: Seven ways to think like a 21st century economist* (1st ed.). Random House Business Books.

Reinecke, J., Manning, S., & von Hagen, O. (2012). The emergence of a standards market: Multiplicity of sustainability standards in the global coffee industry. *Organization Studies*, *33*(5–6), 791–814. https://doi.org/10.1177/0170840612443629

Richardson, R., Steffen, S., Lucht, L., Bendtsen, B., Cornell, C., Donges, D., Drüke Drüke, Fetzer, F., Bala, B., von Bloh, B., Feulner, F., Fiedler, F., Gerten, G., Gleeson, G., Hofmann, H., Huiskamp, H., Kummu, K., Mohan, M., Nogués-Bravo, D., ... Rockström, J. (2023). Earth beyond six of nine planetary boundaries. *Science Advances*, *9*(37), eadh2458. https://doi.org/10.1126/sciadv.adh2458

Ritzer, G., & Dean, P. (2015). *Globalization: A basic text*. John Wiley & Sons, Incorporated.

Sanders, B. (2021, March 29). The rich-poor gap in America is obscene. So let's fix it – here's how. *The Guardian*. https://www.theguardian.com/commentisfree/2021/mar/29/rich-poor-gap-wealth-inequality-bernie-sanders

San Martín, W., & Wood, N. (2021). Pluralising planetary justice beyond the North-South divide: Recentring procedural, epistemic, and recognition-based justice in earth-systems governance. *Environmental Science & Policy*, *128*, 256. https://doi.org/10.1016/j.envsci.2021.12.002

Steffen, W., Richardson, K., Johan Rockström, Cornell, S. E., Fetzer, I., Bennett, E. M., Biggs, R., Carpenter, S. R., Wim, d. V., Wit, C. A. d., Folke, C., Gerten, D.,

Heinke, J., Mace, G. M., Persson, L. M., Ramanathan, V., Reyers, B., & Sverker Sörlin. (2015). Planetary boundaries: Guiding human development on a changing planet. *Science*, *347*(6223), 1259855. https://doi.org/10.1126/science.1259855

Sze, J. (2020). *Environmental justice in a moment of danger*. University of California Press.

Sørensen, E. (2002). Democratic theory and network governance. *Administrative Theory & Praxis*, *24*(4), 693–720. https://doi.org/10.1080/10841806.2002.11029383

Turnpenny, J., Lorenzoni, I., & Jones, M. (2009). Noisy and definitely not normal: Responding to wicked issues in the environment, energy and health. *Environmental Science & Policy*, *12*(3), 347–358. https://doi.org/10.1016/j.envsci.2009.01.004

The Paris Agreement, COP21Cong. (2016). https://unfccc.int/documents/184656

Voß, J. P., & Kemp, R. (2006). Sustainability and reflexive government: Introduction. In *Reflexive governance for sustainable development*. Edward Elgar Publishing. https://doi.org/10.4337/9781847200266.00009

Voulvoulis, N., Giakoumis, T., Hunt, C., Kioupi, V., Petrou, N., Souliotis, I., Vaghela, C., & Wan Rosely, W. (2022). Systems thinking as a paradigm shift for sustainability transformation. *Global Environmental Change*, *75*, 102544. https://doi.org/10.1016/j.gloenvcha.2022.102544

Weaver, M. (2017, January 17). What a pickle: UK gripped by courgette shortage. *The Guardian*. https://www.theguardian.com/lifeandstyle/2017/jan/17/uk-supermarket-courgette-shortage-italy-spain-cold

WEF. (2025). *The global risks report 2025 20th edition*. World Economic Forum. https://reports.weforum.org/docs/WEF_Global_Risks_Report_2025.pdf

3 Sustainable development as transformation

Learning objectives

- Why do we need a societal transformation for sustainable development?
- What does a shift from extractive to regenerative relationship with the planet mean and how does it contribute to sustainable development?
- How are different forms of justice, such as distributive, procedural and recognition justice, relevant to sustainable development?
- What is the green economy paradigm and how does it aim to deliver on sustainable development?
- What alternative economic paradigms are there and how can they support a transformation for sustainable development?

3.1 Introduction

Sustainable development is defined as 'development that meets the needs of the present without compromising the ability of future generations to meet their own needs' (WCED, 1987). Established by the World Commission on Environment and Development (WCED) in 1987, this definition expresses a commitment to equitable needs fulfilment across the globe and obliges us to consider the impact of our actions into the future. However, the 21st-century crises context implies a failure to realise sustainable development in the four decades since its inception. Many have blamed this on the ambiguity of what sustainable development actually means in practice (Redclift, 2005). Since 2015, the United Nations' 17 Sustainable Development Goals (SDGs, Box 3.1) have provided a globally agreed detailed set of targets, sub-targets and indicators for sustainable development to address this ambiguity (UN General Assembly, 2015).

Considering the diverse and unequal global context (Chapter 2), priorities for sustainable development are likely to vary. In parts of the Global South, SDGs 1–9 are urgent as there remains acute need for better access to livelihoods, education, health care and, subsequently, opportunities to thrive. But there is a globally declining trend in almost all indicators of environmental health (Richardson et al., 2023). Environmental degradation from climate change to chemical pollution is impacting our everyday lives regardless of

DOI: 10.4324/9781003179207-4

geographical location, corroding the basis of livelihoods and wellbeing. For example, the international food system discussed in Chapter 6, is becoming destabilised by climate change, loss of soil productivity and genetic diversity, posing a challenge that no country in the world is immune to (Horner, 2020). While goals 1–9 may require urgent attention in low-income countries, it is important that they are not addressed at the expense of goals 10–15.

In the so-called developed countries, urgent emphasis needs to be placed on changing the consumption-heavy lifestyles that drive the planetary crisis, prioritising SDGs 10–15 (Anderson, 2023). That said, inequalities are emerging within developed countries as well. In the UK, for example, the divide between the prosperous south and the lagging north marks significant differences in opportunity and quality of life between regions (HM Government, 2022). Poverty, marginalisation and access to fair work therefore remain relevant concerns across the globe (Graham et al., 2020). A transition away from unsustainable production and consumption practices needs to happen in a just manner, while addressing existing disadvantages and without producing new ones (McCauley & Heffron, 2018). Good governance is therefore central to sustainable development, which requires concerted efforts throughout the globe. Goals 16 and 17 set principles and targets for good governance and international collaboration.

Box 3.1: The UN Sustainable Development Goals (SDGs) as outlined in the Agenda 2030 (UN General Assembly, 2015)

Goal 1. End poverty in all its forms everywhere
Goal 2. End hunger, achieve food security and improved nutrition and promote sustainable agriculture
Goal 3. Ensure healthy lives and promote well-being for all at all ages
Goal 4. Ensure inclusive and equitable quality education and promote lifelong learning opportunities for all
Goal 5. Achieve gender equality and empower all women and girls
Goal 6. Ensure availability and sustainable management of water and sanitation for all
Goal 7. Ensure access to affordable, reliable, sustainable and modern energy for all
Goal 8. Promote sustained, inclusive and sustainable economic growth, full and productive employment and decent work for all
Goal 9. Build resilient infrastructure, promote inclusive and sustainable industrialization and foster innovation
Goal 10. Reduce inequality within and among countries
Goal 11. Make cities and human settlements inclusive, safe, resilient and sustainable
Goal 12. Ensure sustainable consumption and production patterns
Goal 13. Take urgent action to combat climate change and its impacts

Goal 14. Conserve and sustainably use the oceans, seas and marine resources for sustainable development
Goal 15. Protect, restore and promote sustainable use of terrestrial ecosystems, sustainably manage forests, combat desertification, and halt and reverse land degradation and halt biodiversity loss
Goal 16. Promote peaceful and inclusive societies for sustainable development, provide access to justice for all and build effective, accountable and inclusive institutions at all levels
Goal 17. Strengthen the means of implementation and revitalize the Global Partnership for Sustainable Development

The UN SDGs consolidate sustainable development as a global priority, but the way it should be obtained – the route to sustainable development – remains subject to debate. The idea that development is driven by economic growth has prevailed from the era of industrialisation and colonisation (Clark & Szerszynski, 2020; Ghosh, 2021; Jackson, 2025). In the prevalent global liberal capitalist market economy, Gross Domestic Product (GDP – the value of all goods and services produced in a country per annum) holds a powerful position as a key indicator of development (WEF, 2021). This perspective also informs the SDGs, and most explicitly, Goal 8. However, many argue that perpetual economic growth necessitates the kind of extraction of labour and natural resources that cannot be sustained on a finite planet (Folke et al., 2021; Liegey & Nelson, 2020).

As discussed in Chapter 2, if left unregulated, the current financial system and the global free-market economy act as drivers of inequality. Inequality in income and wealth is growing, and the divide between previously colonised countries and the colonising nations has persisted throughout the era of modernisation and continues to do so due to the operations of comparative advantage and subsequent concentration of capital in the global market economy (Ding, 2014; Piketty, 2014). Alternative indicators such as the Genuine Development Index (GDI) suggest that the level of economic activity reached in the 1980s would have been sufficient to achieve wellbeing for all across the globe, had it been fairly and equally distributed, and that any economic growth since has benefited merely a small fraction of the world's population (Jackson, 2017).

3.2 From three pillars to transformation: Rethinking environmental, social, and economic sustainability

Conventionally, sustainable development is seen to require balancing the health of the three dimensions of environment, society and economy (Redclift, 2005; Sachs & Ki-moon, 2015). But in practice, the dimensions are intertwined in a complex manner, as extractive economic growth drives environmental degradation and inequality, and environmental challenges act as poverty and economic risk accelerators (Oxfam, 2023; Piketty, 2014). Many therefore argue that the three dimensions cannot be meaningfully addressed separately and call for a more integrated approach to sustainable development (Scoones et al., 2020; Voulvoulis et al., 2022; Zinecker et al., 2018). This section reviews the notion of sustainability from an environmental (planetary) and socio-economic perspective as a response to both the planetary crisis and the crisis of inequality.

3.2.1 Sustainability and the planetary crisis: From extraction to regeneration

Despite multiple international environmental treaties, domestic legislation and corporate policies, the damage to the environment caused by human activities has continued to escalate in the first decades of the 21st century (Richardson et al., 2023). Given that all life is dependent on the health of the planet (Folke et al., 2021), this 'implementation gap' (Biermann & Kim, 2020) is ultimately self-destructive. The discipline of Earth system science (Folke et al., 2021), explores the complexity of the human-nature relationship. It represents the Earth as a complex dynamic planetary system consisting of different sub-systems such as the atmosphere regulating the climate (Chapter 5) and the land system that is central to food security (Chapter 6). It portrays humans as one species among many, exerting a disproportionate influence on the planetary equilibrium and driving a transition from the stable climate of the Holocene to the high-risk era of the Anthropocene (Chapter 2). The nine Earth system boundaries depicted in Figure 3.1 (see also Chapter 2, Table 2.1) represent the key dynamic processes that are crucial to maintaining the equilibrium of the planetary system and thereby stable living conditions for humans and other living organisms (Steffen et al., 2015).

Scientists have recently determined that we have crossed six of these nine system boundaries, with a high risk of unprecedented changes and disequilibrium in the Earth system (Richardson et al., 2023). In other words, as depicted in Figure 3.1. we are no longer in a safe operating space when it comes to land system change, biosphere integrity (particularly species diversity), climate change, biochemical flows, freshwater change and novel entities (referring to manufactured chemicals) (Stockholm Resilience Centre, 2025). For example, Box 2.2 in Chapter 2 explains how biodiversity loss underpins a collective risk to food security and Box 3.2 below unpacks the role of biochemical flows that support our food system and indeed all primary production on the planet, and the risks that their depletion poses.

As the examples in Boxes 2.2 and 3.2 demonstrate, our economic system is dependent on the continued ability of the Earth system to maintain

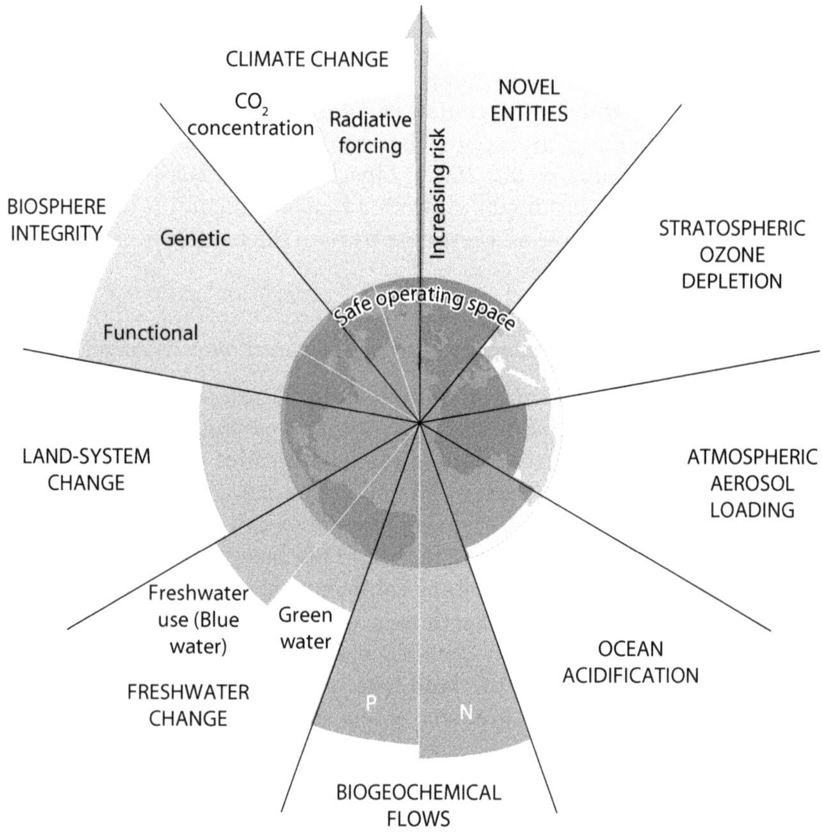

Figure 3.1 The Earth and the nine Earth system boundaries, six of which are now in the zone of high risk of unprecedented developments.

Adapted from Stockholm Resilience Centre, 2025. Licensed under CC BY-NC-ND 3.0. Credit: 'Azote for Stockholm Resilience Centre, based on analysis in (Sakschewski et al., 2025)'.

Box 3.2: The Earth System Boundary presented by biochemical flows

The Earth System Boundary pertaining to biochemical flows refers to the cycling of two of the key nutrients for growth, nitrogen (N) and phosphorus (P), on the planet. These substances are crucial for all living matter and are obtained by plants, animals and people from soil, water and air. But thanks to intensive modern agriculture, these crucial nutrients are overexploited mainly by intensive agricultural practices typical in developed countries such as the US and Australia. Farmers add manufactured fertilisers and/or manure from livestock containing both of these nutrients onto fields to increase plant productivity and crop yields. This often leads to an oversupply of these nutrients in the soil (Ng et al., 2016). When agricultural soil is oversaturated with nitrates, they will either evaporate into the air (as NOx and ammonia), run off into surface watercourses while bound into soil particles that are transported by rainwater or wind, and leach through the soil into groundwater reserves. Within these mediums, they will have a variety of negative impacts, such as eutrophication – excessive growth of living matter in watercourses leading to fish deaths due to the lack of oxygen, and blooms of poisonous algae that hamper recreational uses of watercourses and pose a health threat to both humans and animals. In the air, NOx emitted from agricultural soil, industrial processes and traffic constitute a 'greenhouse gas' responsible for climate change and a health threat (Izcara Palacios, 1998). Nitrates, once inhaled or ingested in water, are associated with multiple human health issues such as asthma and cancers, both of which are on the rise in the human population. Moreover, once nitrates and phosphates are in watercourses, they will eventually end up in oceans, where they are difficult or impossible to capture and return to the 'natural' flow of biochemicals on the planet.

Finally, typically for any sustainability issue, this one is characterised by inequality, underpinned by unequal power relations. While this excessive accumulation of nitrates and phosphates in the 'wrong' place is happening in the regions where intensive mechanised farming is practised, elsewhere, mostly in developing countries, farmers struggle to access these nutrients that would help them maintain the productive capacity of their soils. Therefore, as with many sustainable development issues, the biochemical flows boundary pits the rich against the poor, with responsibility for the problem resting at the feet of large-scale intensive, mechanised, and profit-rich farmers in the global North and the agro-chemicals industry that feeds this overconsumption. Meanwhile, poor nations struggle with the impacts of climate change to which nutrient overapplication elsewhere contributes, and what is more, communities in poor countries suffer from food insecurity due to low crop yields from low-quality soil lacking nutrients.

equilibrium. Climate change offers a further example of the cost of disrupting this ability. In 2024, the Earth has crossed the threshold of 1.5°C of warming, and the impacts of climate change are being felt across the globe (King & Cassidy, 2025). Intense rainfall and flooding are causing damage to settlements and infrastructure and subsequent economic losses, particularly in Asia but also in Europe (King & Cassidy, 2025). Increased drought has caused deforestation and crop losses in semi-arid and arid areas, particularly in the African continent, affecting food prices globally.

The 2022 report by the International Panel on Climate Change (IPCC, 2022) estimates that costs for maintenance and reconstruction of urban infrastructure, including buildings, transportation, and energy, will continue to increase, and the associated disruptions to settlements and infrastructure will be particularly costly in cities. These events also cause human tragedy in the form of loss of homes, territories, and increased mortality. Climate change affects both human physical and mental health, with climate anxiety on the rise among young people (Whitmarsh et al., 2022). Rising et al. (2022) estimate the cost of climate change to the UK economy to be 7.4% of the GDP by 2100.

The foundational role of the planetary system in the economy is often lost due to cultural perceptions of humans as separate from and in control of nature (Jackson, 2025). Science, technological progress, and manufactured infrastructure that we surround ourselves with help uphold this illusion of detachment and control. But underpinning almost all economic activities are so-called 'ecosystem services' (MEA Board, 2005) derived from nature. All of our energy comes from natural systems, and so do our food and raw materials such as minerals, wood, and water. Access to ecosystems even regulates our mood and stress levels. Ecosystems and their services underpin the stability and peacefulness of our societies, for example, by providing sufficient resources to avoid conflicts and mass displacement of communities, as depicted in Figure 3.2.

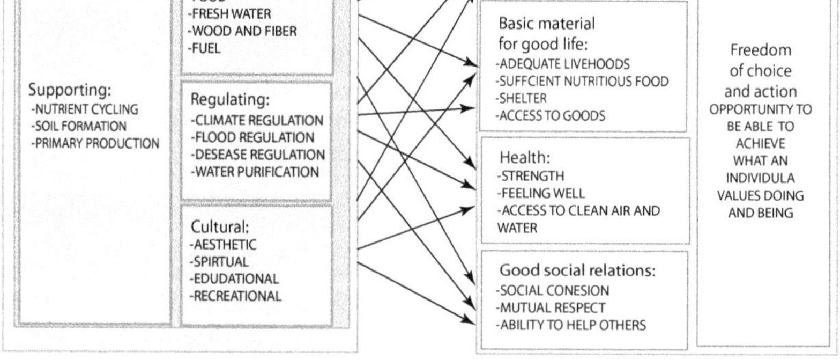

Figure 3.2 The Millennium Ecosystem Assessment classification of ecosystem services and links to the constituents of wellbeing.

Adapted from MEA Board 2005: 7.

This profound dependence of all life forms and with that of human societies and economies on the Earth system is only beginning to be recognised in public and private sector decision-making (Dasgupta, 2021). One way of doing this is to consider nature as a collective of 'assets', which take on significant monetary value in our economies, functioning as 'natural capital' alongside manufactured capital and labour (Dasgupta, 2021; Natural Capital Committee, 2020; UNEP & CBD, 2022). To support this recognition of nature's functions as hugely valuable for humanity, the Millennium Ecosystem Assessment (MEA) (MEA Board, 2005) identified four groups of 'ecosystem services' that 'flow' from 'natural capital' – ecosystems and their functions. The MEA sets out the links between the four groups of services and human wellbeing. This groundbreaking analysis, represented in Figure 3.2, explicates how dependent societies are on nature and its equilibrium.

Sustainability therefore stipulates a cultural change acknowledging our profound dependence on nature and the planetary system. As with any finite resource, to sustain the planet, it is necessary to ensure that we maintain, restore and improve its resources and functions instead of just extracting value from them (Jackson, 2025). Professor Tim Jackson (2017, 2025) suggests that this focus on care needs to replace growth as the central guiding principle of our economic system. In terms of the planetary crisis, this would mean rewarding activities that focus on safeguarding all nine Earth system boundaries and devaluing those that destabilise them. This would imply, for example, a transformation in our relationship with the climate system, from regarding the atmosphere as a site for the assimilation of the by-products of our energy consumption (GHGs); we need to see it as a complex system whose equilibrium we need to work to maintain.

Similarly, in order to safeguard the biosphere, we need to care for the ecosystem services underpinning habitat provision and biodiversity rather than aiming to extract maximum value from them (Dasgupta, 2021). Here, an ethics of care implies a focus on not just protection, but restoration of nature, bringing us back into a 'safe zone' in terms of biosphere integrity and land system change (Richardson et al., 2023). Figure 3.3 depicts the four stages of nature restoration encompassing: (1) reduction of negative environmental and societal impacts, such as pollution and unsustainable resource use and management; (2) removal of contaminants, pollutants and other threats, often known as remediation; (3) rehabilitation of ecosystem functions and services in highly modified areas such as former mining sites and degraded production systems; and (4) ecological restoration, which aims to remove degradation and assists in recovering an ecosystem to the trajectory it would be on if degradation had not occurred, accounting for environmental change (Nelson et al., 2024).

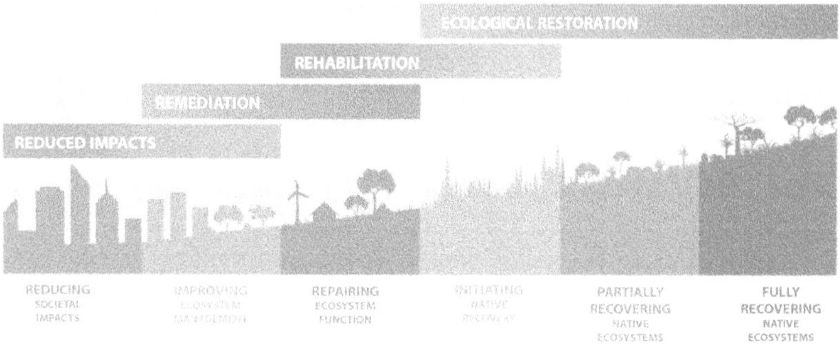

Figure 3.3 The restorative continuum and the four major categories of activities pertaining to the restoration of nature and its ecosystems and biodiversity.

Adapted from Gann et al., 2019.

Regeneration (renewal, restoration or re-growth) and care (maintenance and repair) are therefore central to planetary sustainability (Jackson, 2025). There are emerging examples of regenerative governance interventions. One such example is the 'biodiversity net gain' requirement stipulated by the Environment Act 2021 in the UK. Biodiversity net gain is a development policy designed to ensure that new developments – such as housing provision and transport and energy infrastructure – yield a measurable improvement in biodiversity compared to the pre-development value of the site (+10%) (GFI, 2024). The developer has to ensure that they leave the site in a better condition than it was prior to development and need to sign up to a 30-year management plan to maintain that improved state.

3.2.2 Sustainability as equity and social justice

Sustainable development is not just about caring for the environment but needs to also focus on people's ability for needs fulfilment (WCED, 1987). The UN SDGs denote addressing poverty, hunger and access to resources as central sustainable development concerns. At the time of writing, the World Bank's poverty line defines anyone with an income of less than US $3 per day as living in extreme poverty. In 2025, 831 million people were living in poverty according to this metric, which constitutes approximately 10% of the global population (World Bank Group, 2025). Monetary indicators of poverty are widely used but ultimately, extreme poverty can be understood as inability to meet basic material needs for food, clean water, sanitation, shelter, health care, education and public services such as transport and energy provision (Sachs & Ki-moon, 2015).

Despite the commitment to end poverty by 2030 (SDG1), the proportion of people living in poverty globally has hovered between 8 and 10% in the years from 2018 to 2024 (United Nations Department of Economic and

Social Affairs, 2025). Moreover, some parts of the world can be described as stuck in extreme poverty. This is the case, for example, in Sub-Saharan Africa, which accommodates the majority of the world's population living in extreme poverty (over 464 million people), while poverty rates in China and the rest of East Asia have declined in recent decades (United Nations Department of Economic and Social Affairs, 2025).

The mechanisms barring certain households and countries from benefitting from the ongoing global GDP growth were outlined in Chapter 2 and can be summarised as the pressures for competitiveness and efficiency driven by the global free market system, which compels governments and businesses to increasingly efficient value extraction from labour, capital and natural resources; the lack of robust redistribution mechanisms that would address the accumulation of income and wealth in the hands of capital owners; and the biased rules of the global financial system that favour the already rich countries that were instrumental in setting them (Chancel et al., 2026).

As discussed above, the planetary crisis acts as an accelerator of poverty and disadvantage. Globally, vulnerability to the impacts of climate change is concentrated in West, Central and East Africa, South Asia, Central and South America, Small Island Developing States and the Arctic. These are also locations where poverty, weak governance and limited access to basic services and resources are prevalent (IPCC, 2022). Poor countries where livelihoods like farming account for a larger proportion of income are directly vulnerable to the impacts of climate change. The ability to adapt to the changing climate is lower due to limited scope for investment in technology and lack of socio-economic safety nets such as insurance and benefit schemes. IPCC's 2022 report on the impacts of climate change states that between 2010 and 2020, human mortality from floods, droughts and storms was 15 times higher in highly vulnerable regions, compared to regions with very low vulnerability. Vulnerability is exacerbated at the local scale by inequity and marginalisation linked to gender, ethnicity and poverty.

Sustainable development must address both socio-economic inequality and inequality in terms of the impact of the planetary crisis on people and communities. To meaningfully address inequality beyond needs fulfilment, it is necessary to understand the premises of healthy communities and what enables people to thrive. SDGs 1–9 suggest that, beyond basic needs for nutrition and health, access to education, sanitation and gender equality are important. The International Association for Impact Assessment lists the following variables as central to the wellbeing of people and communities (Vanclay, 2003):

- *People's way of life* – ability to live, work, play and interact with one another;
- *Their culture* – the integrity of shared beliefs, customs, values, and language or dialect;
- *Their community* – its cohesion, stability, character, services and facilities;

- *Their political systems* – the extent to which people are able to participate in decisions that affect their lives, the level of democratisation that is taking place, and the resources provided for this purpose;
- *Their environment* – the quality of the air and water people use; the availability and quality of the food they eat; the level of hazard or risk, dust and noise they are exposed to; the adequacy of sanitation, their physical safety, and their access to and control over resources;
- *Their health and wellbeing* – where health is a state of complete physical, mental, social and spiritual wellbeing and not merely the absence of disease or infirmity;
- *Their personal and property rights* – particularly whether people are economically affected, or experience personal disadvantage which may include a violation of their civil liberties; and
- *Their fears and aspirations* – their perceptions about their safety, their fears about the future of their community, and their aspirations for their future and the future of their children.

The social dimension of sustainable development focuses on socio-economic inequality, vulnerability and broader social value, indicated by the above variables. A multidimensional understanding of justice is therefore central to sustainable development (Kaljonen et al., 2021; Redclift, 2005; Sachs & Ki-moon, 2015). Distributional justice focuses on the distribution of income and wealth but also of impacts from environmental harms and their remedial actions on people's way of life, property rights, jobs and fears and aspirations (Murdoch, 2020). Taking communities and other stakeholders into consideration requires understanding their varied perspectives, interests and experience, which are best captured if stakeholders are allowed to participate in governance (Dendena & Cordi, 2015). Procedural justice refers to the scope for participation – providing all those affected a fair chance of taking part or being represented in decision-making about issues that affect them (Murdoch, 2020). Procedural justice safeguards the ability of people to influence decision making regarding issues of common interest.

A third dimension of justice that is increasingly seen as necessary for integrating plural perspectives and values into governance is termed recognition justice (Coolsaet & Neron, 2020). It pertains to the experiences of marginalised groups whose perspectives may differ quite radically from those that are routinely acknowledged as stakeholders in decision-making. Recognition justice speaks directly to the crisis of knowledge in that it implicates the often unspoken rules about valid knowledge and valid ways of participation and agitates the inherent bias in the dominant Western science-informed perspective on development (Ghosh, 2021; Gupta et al., 2023).

For example, overreliance on Western scientific knowledge has led to the marginalisation of Indigenous perspectives which underpin demonstrably better approaches to biodiversity provision (Lélé, 1991). The most notable shortcoming of Western notions of wellbeing and development is the perception of humans and nature as separate entities. This ignores our fundamental dependence on nature and departs from most Indigenous perspectives and notions of good life (Kothari et al., 2014).

Global environmental justice alerts us to the differing responsibility for the planetary crisis globally and calls for remedial actions (IPCC, 2022; Ritchie & Roser, 2024). All the significant environmental problems represented by the Earth system boundaries, such as climate change, biodiversity loss and the disruption of biochemical flows are caused by activities associated with so-called developed nations and extractive economic growth (Gómez-Baggethun & Naredo, 2015). For example, the majority of greenhouse gas emissions stem from rich countries. Viewing cumulative emissions, the biggest emitters are the USA and Europe, which were also responsible for the biggest emissions per annum well into the 21st century (Ritchie et al., 2020). Although Asia and in particular, China and India have recently joined the USA as the top three emitters of carbon per annum, the historic and per capita emissions in India are much smaller than in the US or China, or indeed in most European countries.

When focusing on the emissions associated with consumption of goods and services by the people living in a country, the responsibility rests even more explicitly with developed nations and their populations (Ritchie et al., 2020). A large proportion of China's emissions stem from the production of goods consumed in rich countries. Environmental justice calls for further justice perspectives such as reparative justice, referring to the need to compensate for past losses, such as damage in developing countries due to climate change caused by developed countries. Cosmo justice or planetary justice expands this to integrate non-human species and natural entities into the scope of stakeholders who have rights and deserve justice (Gupta et al., 2023).

Sustainable development must address multiple complex drivers and manifestations of inequality. There is mounting evidence that the current growth-oriented economic system reinforces a harmful and unfair extractive approach to Earth's resources and functions (Raworth, 2017; Rigon, 2017). This points to a need to rigorously interrogate the role of the growth-oriented liberal market economy in the 21st-century crises and carefully consider its role in sustainable development.

3.3 Sustainable development: Green economy and societal transformation

This section contrasts two different approaches to realising sustainable development in the 21st-century context and examines the role of the global economic system in the multi-crises in more detail.

3.3.1 Green economy

Today still, mainstream economics and policy subscribe to the theory that economic growth needs to be prioritised to support healthy societies and the planet. This has given rise to the notion of the green economy, which assumes that economic activity can be steered, with governance interventions and consumer demand, towards activities that sustain the planet (Gómez-Baggethun & Naredo, 2015). The idea that nature should be assigned economic value, as natural capital discussed above, aligns with the green economy paradigm. Many governance interventions integrating the value of nature into economic processes in various ways are already in place. For example, subsidies encouraging agricultural practices that safeguard biodiversity, and higher taxes for more polluting cars that use up more of the athmosphere's capacity to assimilate waste (Evans & Thomas, 2024). The green economy frames the necessary shift to safeguard planetary sustainability as vested with economic opportunities, resulting for example from the need for new renewable energy infrastructure and associated jobs (HM Government, 2019). The notion of the circular economy, where recycling significantly decreases the need for raw materials and the amount of waste, is an iteration of the green economy paradigm (Chirambo, 2021).

In a green economy, the need for natural resources such as minerals and water is reduced through the adoption of more efficient technologies, recycling of waste and extended lifespan of consumer goods or by replacing polluting economic activities with non-polluting ones (Chirambo, 2021; Keyßer & Lenzen, 2021). It is therefore based on the notion of 'decoupling' economic growth from environmental resource use and pollution (Figure 3.4 and Haberl et al., 2020). Haberl et al. (2020) refer to 'absolute decoupling' where resource use and pollution decline while the economy keeps growing, and 'relative decoupling' where resource use and emissions are still growing but at a lower rate than the economy (Figure 3.4).

Both in the US and Europe, the green economy paradigm has translated into policy initiatives that seek to render environmental sustainability into a profitable business that provides jobs, investment opportunities and above all, contributes positively to the GDP. The European 'Green Economy Policy Package' and the American 'Green New Deal' (UNEP, 2011) seek profit from economic activities like retrofitting buildings to render them carbon neutral and look to the renewable energy industry to offer new jobs and provide a market for new lucrative innovation and investment opportunities (Jackson, 2017).

The green economy agenda is also reflected in the UN SDGs, where SDG 8 calls for sustainable economic growth through innovation and efficiency gains, while also paying heed to labour rights, ending modern slavery and provision of equal opportunity (UN General Assembly, 2015; United Nations Department of Economic and Social Affairs, 2025). Chirambo (2021) suggests that particularly goals 7 (access to clean energy) and 13 (climate action) need to attract venture capital (international investment from

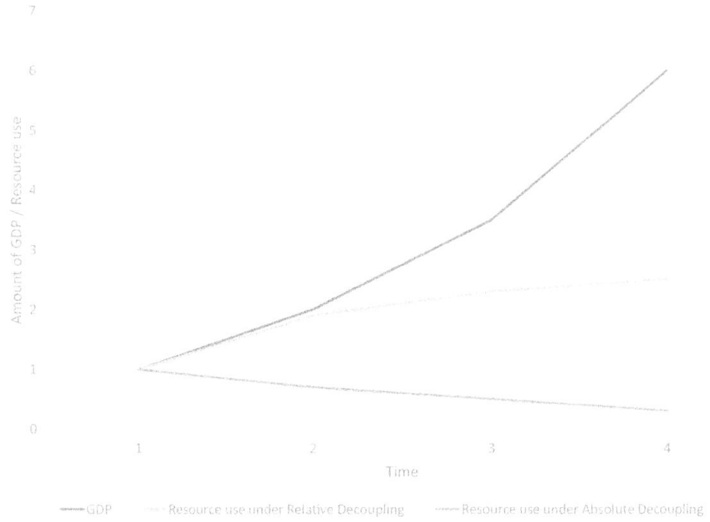

Figure 3.4 Relative decoupling of resource extraction from economic growth (GDP line) happens when GDP grows faster than resource use. Absolute decoupling, on the other hand, implies a negative correlation between resource extraction and GDP.

both private and institutional funders) to fund so-called 'climate change entrepreneurship'. Both governments and the private sector have been marshalled to invest in the SDGs. For example, in Japan, the goals have inspired the technology-driven 'Green Transformation' strategy advocating for a 'super smart society' focusing on innovations that add value but operate in harmony with nature (Bosquerolli, 2025). A key part of the Japanese SDG strategy is a charter of corporate behaviour for the realisation of a sustainable society.

The green economy paradigm is therefore thriving in implementation and seems to have gained sufficient traction to remain relatively unaffected by fluctuations in high-level political support for the green agenda. 'Relative decoupling' (Figure 3.4), where resource use and emissions are growing at a lower rate than the economic contribution, is becoming more frequent (Haberl et al., 2020). In high-income countries, there are examples of absolute decoupling of GDP and CO_2 emissions mostly due to energy efficiency improvements, a shift to renewables and outsourcing of carbon-intensive production. However, at the global scale, absolute decoupling of GDP growth and CO_2 emissions remains absent. This is problematic, suggesting that the green economy paradigm alone is insufficient to reach crucial international targets for sustainable development (such as SDG 13 and the Paris Agreement on limiting Climate Change to 2°C) (Haberl et al., 2020; Infante-Amate et al., 2024). Haberl et al. (2020) emphasise the need for sufficiency-oriented strategies (reducing overall consumption and production) and strict enforcement of absolute reduction targets for resource use and emissions.

Further challenges include a demonstrable 'sustainability skills gap' globally, where a range of policies are now pushing for green technologies, green solutions and green policies, but 'green expertise' to actually deliver on this remains thin on the ground (Calder & Drayson, 2016). Moreover, the political will to set high enough incentives to reach environmental targets, for example, through taxation of polluting activities and goods and cap-and-trade systems setting overall limits to emissions is absent (Evans & Thomas, 2024). Given the urgency of the planetary crisis and how much climate change alone is costing in terms of monetary losses and human suffering, critics suggest that the green economy can at best deliver incremental improvements. Calls for more fundamental changes and a prominent critique of the economic growth imperative and the current dominant economic system are emerging within sustainable development literature. The next section explores some of the alternative iterations of sustainable development that propose new ways of conceptualising and achieving wellbeing and opportunity to thrive for all.

3.3.2 Sustainable development as transformation: Emerging contenders

The lack of success in decoupling resource use from the growth of economic activity suggests that a change in the logic of the economic system may be necessary to address the planetary crisis. Many suggest that the prevalent extractive rationale needs to be replaced by a system that rewards a regenerative approach. Sensitive to the finite capacity of the planetary system. Moreover, the green economy paradigm does not address the growing socio-economic polarisation. In fact, it is firmly anchored within existing economic relations, dominant Western knowledge claims, and technological solutions that constitute the drivers of many of the 21st-century crises. In order to deliver on both planetary and social sustainability, sustainable development needs to mark not just a shift but a transformation in the prevalent structures and processes shaping our relationship with the planet (Scoones et al., 2020). This needs to encompass the way humans engage with the planetary resources and functions and instil an alternative economic model that decentres growth in favour of planetary and human wellbeing as an indicator of success (Clark & Szerszynski, 2020; Gupta et al., 2023; Jackson, 2025; Liegey & Nelson, 2020).

In her thesis on Doughnut Economics (Chapter 1), Kate Raworth argues that sustainable development should aim all efforts at reaching 'a safe and just space for humanity' by means of a sustainable 'regenerative and distributive economy' (Raworth, 2017). This safe and just space – the doughnut – sits within the pie-chart image of the nine Earth system boundaries (Figure 3.5), where its outer boundary aligns with those of the Earth system and its inner boundary – the hole of the doughnut, circles a space of 'critical human deprivation'. The doughnut itself marks the space that is 'safe and just' for all

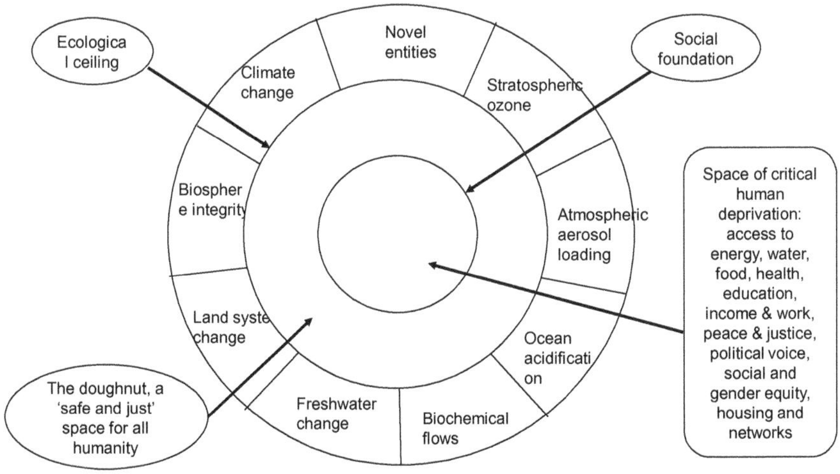

Figure 3.5 The doughnut representing a safe and just space for humanity.
Adapted from: Raworth 2017.

humanity, situated above the 'social foundation' where critical deprivation is avoided and below the 'ecological ceiling' which marks drift into resource use that risks the functioning of the Earth system. Raworth suggests that instead of economic growth, access to water, food, health, education, peace and justice, political voice, social equity, gender equality, housing, networks and energy should be used as the key indicators of development, with the aim of reaching the doughnut on all of them without overshooting the ecological ceiling.

But Raworth's depiction of sustainable development as an effort to remain in the safe zone of the doughnut does not tell us much about how exactly we are to reach that zone. An emerging literature on various post- or degrowth trajectories (Kallis et al., 2020; Liegey & Nelson, 2020) that decentre economic growth articulates options for delivering wellbeing on a planet with finite resources and capacities. Ashish Kothari et al. (2014) point to the need to look beyond the Western perspective to principles such as 'care', 'conviviality', 'sharing', 'simplicity', and the 'commons' as key to sustainable and equitable economies. This offers a vocabulary and an imagery that replaces perpetual economic growth and the commodification of nature as necessary for sustainable development (Kothari et al., 2014).

A sustainable degrowth economy prioritises and rewards activities and targets that maintain and repair the planet and that foster diversity and conviviality, which are central to resilience and wellbeing (Jackson, 2025). Hickel (2020) defines degrowth as 'a planned reduction of excess energy and resource use to bring the economy back into balance with the living world in a safe, just and equitable way'. This requires a transformation in currently dominant values and social and economic practices. For example, the principle of care troubles prevalent values that centre on consumerism. Where polluting and resource-intensive items and habits such as large cars and long-haul flights are currently seen as positive status symbols, we need to become sensitive to the excessive consumption of resources and lifestyles lacking in 'care' for the planet and other humans that these practices represent. This change in perspective could facilitate the necessary reduction of material and energy consumption to address the planetary crisis (Martínez-Alier et al., 2010). The currently popular notion of wellbeing needs to be expanded beyond the self to encompass others, including other species living on the planet. This 'conviviality' would place emphasis on extended social networks and inclusivity as demonstrating success and secure status in society and help to reorient technological development to support planetary wellbeing rather than growth.

In addition to a transformation in values, degrowth implies a substantial transformation of the current global free-market economy. Economic valuation needs to be fundamentally reprogrammed to reward activities that align with these regenerative and equitable values and norms (Liegey & Nelson, 2020). Wages should reward activities that care for the restorative functions of nature and our own health rather than extract from them (Jackson, 2025). To an extent, this borrows from the green growth paradigm. Community organisations that have begun to implement the degrowth idea into practice suggest, in line with the circular economy paradigm, that taxation and pricing of goods should be changed to reflect their shareability, reparability and lifespan (Rigon, 2017). Income and wealth redistribution mechanisms such as taxes and public investment have proven powerful means to address inequality, and these could be used more effectively to incentivise activities that serve the purpose of the safe space for humanity (Raworth, 2017).

Many proponents of degrowth therefore embrace some aspects of the green economy and emphasise the need for green growth particularly in poor countries but not growth for growth's sake and by any means (Hickel, 2020). However, they condemn the tendency to represent nature in monetary terms (as natural capital) and suggest that this risks enhancing its subjugation into the economy and the processes of commodification and consumption that are problematic in themselves (Martínez-Alier et al., 2010; O'Neill, 2017). Instead, nature should be managed as a 'commons', shared property that everyone has a right to access and whose sustainable management is therefore in everyone's interest as safeguarding this right (Hiedanpaa & Bromley, 2016).

Re-localising economic interactions is seen as central to resisting the power of global economic forces and valorising care for local ecosystems and wellbeing. Favouring sustainable small-scale businesses ensures that local livelihoods thrive and can reduce the carbon footprint of goods (Martínez-Alier et al., 2010). This can be encouraged through the introduction of a local currency (such as the Lewes Pound: https://www.thelewespound.org/) that is used alongside a national currency. A local currency based on the exchange of local goods and services can support the emergence of localised non-monetary economies where increased transparency provides accountability for environmentally and socially sustainable products and services (North, 2018). This offers an alternative to people who do not want to support multinational companies and long supply chains characterising the mainstream, growth-oriented economy (see e.g. Chapter 6 for the global food system).

Degrowth scholars and practitioners also advocate reducing the time allocated to work (Hickel, 2020; Liegey & Nelson, 2020). An improvement to the so-called work-life balance can in itself increase wellbeing for many, while simultaneously reducing the pressure to produce and consume resource-intensive goods. Similarly, access to a universal basic income could free people to work towards wellbeing at the local scale rather than as a part of a complex international provisioning system. A guaranteed basic income could further support a re-localisation of food production by enabling those who wish to grow food at a small scale to supplement their own needs and those of the local community. Urban agriculture has been proven to have a myriad of benefits for diets, wellbeing and climate resilience (Huang et al., 2025). Some countries have experimented with basic income, a baseline income guaranteed for everyone by the state, to address persistent unemployment and its detrimental impacts on mental health and to recognise the value of often upaid reproductive labour such as childcare at home (BBC, 2019).

Liegey and Nelson (2020) suggest that degrowth, an actual contraction of econimic activity, may be an inevitable outcome considering the extent to which our present lifestyles and the economic model that supports them are dependent on finite energy sources such as oil and gas. We have already passed 'peak oil' beyond which oil extraction gradually becomes so expensive that it no longer affords economic growth. The post-pandemic trends in oil and gas prices certainly give credence to this argument, as global political developments further exacerbate access to these energy sources. Degrowth pathways arguably present more equitable routes to sustainability than ones relying on novel, green technology solutions such as electrification of transport systems and carbon capture (Keyßer & Lenzen, 2021). But political will to shift focus from GDP to more relevant indicators of wellbeing is distinctively absent presently. Those who benefit from the status quo – even as the vast majority of people are experiencing a decrease in living standards and the quality of life due to economic crises and climate change – are the ones in power presently and they are reluctant to look to alternatives (Oxfam, 2023). A better understanding of the structural, institutional, political, and social barriers to change and the potential unexpected impacts of these alternative development pathways is required.

3.4 Conclusions

The WCSD's definition of 'development that meets the needs of the present without compromising the ability of future generations to meet their needs' (WCED, 1987) sums up the key message of sustainable development. But what the term actually means in practice, and what its key principles are, is in evolution. The SDGs marked the return of sustainable development onto the global agenda. The 17 goals represent globally recognised key concerns and place significant emphasis on equity and shared responsibility for global development, responding to evolving research and policy debates since the launch of the WCED report in 1987.

However, the multi-crises within which we find ourselves in the 21st century call for transformational change in institutions, power relations, values and norms – and in society as a whole. Instead of economic growth, sustainable development needs to prioritise equal access to energy, food, water, health, housing, work, political representation and a clean and meaningful environment while safeguarding the boundaries of the Earth system (Raworth, 2017). This requires *fundamental societal transformation towards equitable and regenerative relations among people and the planet, addressing the core drivers of unsustainability*. The care-economy and degrowth offer pathways towards a more regenerative future through ground-level solutions that could in time be mainstreamed to replace the neoliberal approach stipulated by the global capitalist market economy (Jackson, 2025; Kothari et al., 2014).

Careful thought is needed as to whether and how a shift away from a growth-oriented global economic model impacts developing economies where extreme poverty is still a significant and persistent problem. More robust redistribution mechanisms mitigating the gap between the very rich and the poor and rewarding regenerative rather than extractive activities seem necessary. An emerging literature points to alternative approaches drawing from non-Western philosophies and ways of life as valuable complementary frameworks (Ghosh, 2021; Kothari et al., 2014). Instilling values such as conviviality and care and re-localising economic relations would help resist the processes of marginalisation driven by export-led growth trajectories. Ultimately, growth-oriented policies such as the SDGs need to be replaced by ones that facilitate a shift in the rationale and metrics of sustainable development that observe the Earth system boundaries and surpass the minimum threshold of critical human deprivation by ensuring that the Earth's regenerative capacity is nurtured and everyone's critical needs are met regardless of social status or geographical location.

References

Anderson, K. (2023, March 24). IPCC's conservative nature masks true scale of action needed to avert catastrophic climate change. *The Conversation*. Retrieved September 5, 2025, from http://theconversation.com/ipccs-conservative-nature-masks-true-scale-of-action-needed-to-avert-catastrophic-climate-change-202287

BBC. (2019, February 8). Finland basic income trial left people 'happier but jobless'. https://www.bbc.co.uk/news/world-europe-47169549

Biermann, F., & Kim, R. (2020). Architectures of earth system governance: Setting the stage. In F. Biermann & R. Kim (Eds.), *Architectures of earth system governance: Institutional complexity and structural transformation* (pp. 1–34). Cambridge University Press.

Bosquerolli, A. M. (2025). Evolution of sustainable development in Japan: From historical foundations to green transformation. *Asian Journal of Agricultural*

Extension, Economics & Sociology, *43*(3), 54–65. https://doi.org/10.9734/ajaees/2025/v43i32701
Calder, W., & Drayson, H. (2016). *Student attitudes towards, and skills for, sustainable development & employer attitudes towards, and skills for, sustainable development*. SAGE Publications. https://doi.org/10.1177/0973408216652470
Chancel, L., Gómez-Carrera, R., Moshrif, R., & Piketty, T. (2026). *World inequality report 2026*. https://wid.world/document/world-inequality-report-2026/
Chirambo, D. (2021). Corporate sector policy innovations for Sustainable Development Goals (SDGs) implementation in the Global South: The case of sub-Saharan Africa. *Journal of Sustainability Research*, *3*(2). https://doi.org/10.20900/jsr20210011
Clark, N., & Szerszynski, B. (2020). *Planetary social thought: The anthropocene challenge to the social sciences*. Polity Press.
Coolsaet, B., & Neron, P. (2020). Recognition and environmental justice. In B. Coolsaet (Ed.), *Environmental justice: Key issues* (pp. 52–63). Taylos & Francis Group.
Dasgupta, P. (2021). *The economics of biodiversity: The Dasgupta review abridged version*. HM Treasury.
Dendena, B., & Cordi, S. (2015). The environmental and social impact assessment: A further step towards an integrated assessment process. *Journal of Cleaner Production*, *108*, 956–977.
Ding, C. (2014). The complex social side of globalisation. In M. Telò (Ed.), *Globalisation, multilateralism, Europe: Towards a better global governance?* (pp. 249–267). Taylor & Francis Group.
Evans, J. P., & Thomas, C. (2024). *Environmental governance* (2nd ed.). Routledge.
Folke, C., Polasky, S., Rockström, J., Galaz, V., Westley, F., Lamont, M., Scheffer, M., Österblom, H., Carpenter, S. R., Chapin, F. S., Seto, K. C., Weber, E. U., Crona, B. I., Daily, G. C., Dasgupta, P., Gaffney, O., Gordon, L. J., Hoff, H., Levin, S. A., ... Walker, B. H. (2021). Our future in the anthropocene biosphere. *Ambio*, *50*(4), 834–869. https://doi.org/10.1007/s13280-021-01544-8
Gann, G. D., McDonald, T., Walder, B., Aronson, J., Nelos, C. R., Jonson, J., Hallett, J. G., Eisenberg, C., Guariguata, M., Liu, J., Hua, F., Echeverria, C., Gonzales, E., Shaw, N., Decleer, K., & Dixon, K. W. (2019). International principles and standards for the practice of ecological restoration. Second edition. *Restoration Ecology*, *27*(S1), S1–S46. https://doi.org/10.1111/rec.13035
GFI. (2024). *Biodiversity net gain*. Green Finance Institute. https://hive.greenfinanceinstitute.com/wp-content/uploads/2024/10/R4N-GUIDEBOOKS-BNG-ENGLAND.pdf
Ghosh, A. (2021). *The Nutmeg's Curse parables for a planet in crisis*. university of Chicago Press. https://doi.org/10.7208/chicago/9780226815466
Gómez-Baggethun, E., & Naredo, J. M. (2015). In search of lost time: The rise and fall of limits to growth in international sustainability policy. *Sustainability Science*, *10*(3), 385–395. https://doi.org/10.1007/s11625-015-0308-6
Graham, M., Woodcock, J., Heeks, R., Mungai, P., Van Belle, J., du Toit, D., Fredman, S., Osiki, A., van der Spuy, A., & Silberman, S. M. (2020). The Fairwork Foundation: Strategies for improving platform work in a global context. *Geoforum, 112*, 100–103. https://doi.org/10.1016/j.geoforum.2020.01.023
Gupta, J., Liverman, D., Prodani, K., Aldunce, P., Bai, X., Broadgate, W., Ciobanu, D., Gifford, L., Gordon, C., Hurlbert, M., Inoue, C. Y. A., Jacobson, L., Kanie, N., Lade, S. J., Lenton, T. M., Obura, D., Okereke, C., Otto, I. M., Pereira, L., . . . Verburg, P. H. (2023). Earth system justice needed to identify and live within Earth system boundaries. *Nature Sustainability, 6*(6)10.1038/s41893-023-01064-1
Haberl, H., Wiedenhofer, D., Vir´ag, D., Plank, B., Kalt, G., Brockway, P., Fischman, T., Hausknost, D., Krausmann, F., Leon-Gruchalski, B., Mayer, A., Pichler, M., Schaffartzik, A., Sousa, T., Streeck, J., & Creutzig, F. (2020). A systematic review of the evidence on decoupling of GDP, resource use and GHG emissions, part II: synthesizing the insights. *Environmental Research Letters, 15*(6), 1–42.

Hickel, J. (2020). *Less is More: How Degrowth Will Save the World*. Random House.

Hiedanpaa, J., & Bromley, D. (2016). *Environmental Heresies: The Quest for Reasonable*. Palgrave Macmillan. https://doi.org/10.1057/978-1-137-60083-7

HM Government. (2019). *Voluntary National Review of progress towards the Sustainable Development Goals, United Kingdom of Great Britain and Northern Ireland*. Right Vision Media.

HM Government. (2022). *Levelling Up the United Kingdom*

Horner, R. (2020). Towards a new paradigm of global development? Beyond the limits of international development. *Progress in Human Geography, 44*(3), 415–436. https://doi.org/10.1177/0309132519836158

Huang, I. Y., Forero, O. A., Wagner-Medina, E. V., Florez Diaz, H., Tremma, O., Fargetton, X., & Lowenberg-Deboer, J. (2025). Resilience of food supply systems to sudden shocks: A global review and narrative synthesis. *Global Food Security, 44*10.1016/j.gfs.2025.100842

Infante-Amate, J., Travieso, E., & Aguilera, E. (2024). Unsustainable prosperity? Decoupling wellbeing, economic growth, and greenhouse gas emissions over the past 150 years. *World Development, 184*, 106754. https://doi.org/10.1016/j.worlddev.2024.106754

IPCC. (2022). *Climate Change 2022: Impacts, Adaptation, and Vulnerability. Contribution of Working Group II to the Sixth Assessment Report of the Intergovernmental Panel on Climate Change*. Cambridge: Cambridge University Press. https://www.ipcc.ch/report/sixth-assessment-report-working-group-ii/

Izcara Palacios, S. (1998). Farmers and the Implementation of the EU Nitrates Directive in Spain. *Sociologia Ruralis, 38*(2), 144–162. https://doi.org/10.1111/1467-9523.00069

Jackson, T. (2017). *Prosperity without growth : foundations for the economy of tomorrow* (2nd ed.). Routledge. https://doi.org/10.4324/9781315677453

Jackson, T. (2025). *The care economy*. Polity.

Kaljonen, M., Kortetmäki, T., Tribaldos, T., Huttunen, S., Karttunen, K., Maluf, R. S., Niemi, J., Saarinen, M., Salminen, J., Vaalavuo, M., & Valsta, L. (2021). Justice in transitions: Widening considerations of justice in dietary transition. *Environmental Innovation and Societal Transitions, 40*, 474. https://doi.org/10.1016/j.eist.2021.10.007

Kallis, G., Paulson, S., D'Alisa, G., & F. Demaria. (2020). *The case for Degrowth*. Cambridge: Polity Press.

Keyßer, L. T., & Lenzen, M. (2021). *1.5 °C degrowth scenarios suggest the need for new mitigation pathways*. Springer Science and Business Media LLC. https://doi.org/10.1038/s41467-021-22884-9

King, A., & Cassidy, L. (2025, -02-10). Earth is already shooting through the 1.5°C global warming limit, two major studies show. *The Conversation*. Retrieved September 30, 2025, from http://theconversation.com/earth-is-already-shooting-through-the-1-5-c-global-warming-limit-two-major-studies-show-249133

Kothari, A., Demaria, F., & Acosta, A. (2014). Buen Vivir, Degrowth and Ecological Swaraj: Alternatives to sustainable development and the green economy. *Development (Society for International Development)*, 57(3–4), 362–375. https://doi.org/10.1057/dev.2015.24

Lélé, S. M. (1991). Sustainable development: A critical review. *World Development, 19*(6), 607–621. https://doi.org/10.1016/0305-750X(91)90197-P

Liegey, V., & Nelson, A. (2020). *Exploring Degrowth: A critical guide* (1st ed.). Pluto Press.

Martínez-Alier, J., Pascual, U., Vivien, F., & Zaccai, E. (2010). *Sustainable de-growth: Mapping the context, criticisms and future prospects of an emergent paradigm*. Elsevier BV. https://doi.org/10.1016/j.ecolecon.2010.04.017

McCauley, D., & Heffron, R. (2018). Just transition: Integrating climate, energy and environmental justice. *Energy Policy*, *119*, 1–7. https://doi.org/10.1016/j.enpol.2018.04.014

MEA Board. (2005). *Living beyond our means: Natural assets and human well-being. Statement from the Board*. Millennium Ecosystem Assessment.

Living beyond our means: Natural assets and human wellbeing. (2005).

Murdoch, E. (2020). *A history of environmental justice foundations, narratives, and perspectives* (B. Coolsaet, Ed., pp. 6–17). Routledge Taylor & Francis.

Natural Capital Committee. (2020). *Final response to the 25 year environment plan progress report*. Natural Capital Committee. https://assets.publishing.service.gov.uk/media/5fa95768d3bf7f03acd13a4f/ncc-final-response-25yr-env-plan.pdf

Nelson, C. R., Hallett, J. G., Romero Montoya, A. e., Adrade, A., Besacier, C., Boerger, V., Bouazza, K., Chazdon, R., Cohen-Sacham, E., Danano, D., Diedrichsen, A., Fernandez, Y., Gann, G. D., Gonzales, E. K., Gruca, M., Guariguata, M. R., Gutierrez, V., Hancock, B., Innecken, P., ... Weidlich, E. W. A. (2024). *Standards of practice to guide ecosystem restoration: A contribution to the United Nations Decade on ecosystem restoration 2021–2030*. FAO. https://doi.org/10.4060/cc9106en

Ng, E., Chen, D., & Edis, R. (2016). Nitrogen pollution: The forgotten element of climate change. *The Conversation*. https://theconversation.com/nitrogen-pollution-the-forgotten-element-of-climate-change-69348

North, P. (2018). *Alternative currency movements as a challenge to globalisation? A case study of Manchester's local currency networks*. Routledge.

O'Neill, J. (2017). Life beyond capital | Essay by John O'Neill. https://cusp.ac.uk/themes/nature/m1-6/

Oxfam. (2023). *Climate equality: A planet for the 99%*. Oxfam GB. https://doi.org/10.21201/2023.000001

Pelling, M. (2012). Resilience and transformation. In M. Pelling, D. Navarrete-Manuel, & RedcliftM. (Eds.), *Climate change and the crisis of capitalism*. Routledge.

Piketty, T. (2014). *Capital in the twenty-first century*. Harvard University Press.

Raworth, K. (2017). *Doughnut economics: Seven ways to think like a 21st century economist* (1st ed.). Random House Business Books.

Redclift, M. (2005). Sustainable development (1987–2005): An oxymoron comes of age. *Sustainable Development*, *13*, 212–217. https://doi.org/10.1002/sd.281

Richardson, R., Steffen, S., Lucht, L., Bendtsen, B., Cornell, C., Donges, D., Drüke Drüke, Fetzer, F., Bala, B., von Bloh, v. B., Feulner, F., Fiedler, F., Gerten, G., Gleeson, G., Hofmann, H., Huiskamp, H., Kummu, K., Mohan, M., Nogués-Bravo, D., ... Rockström, J. (2023). Earth beyond six of nine planetary boundaries. *Science Advances*, *9*(37), eadh2458. https://doi.org/10.1126/sciadv.adh2458

Rigon, A. (2017). *Degrowth politics and policies for Degrowth*. Degrowth. Retrieved September 10, 2025, from https://degrowth.info/blog/degrowth-politics-and-policies-for-degrowth

Ritchie, H., & Roser, M. (2024, January 22). *CO_2 emissions. How much CO_2 does the world emit? Which countries emit the most?* Our World in Data. Retrieved March 12, 2024, from https://ourworldindata.org/co$_2$-emissions

Ritchie, H., Roser, M. & Rosado, P. (2020). *CO? and greenhouse gas emissions*. Our World in Data. Retrieved October 28, 2022, from https://ourworldindata.org/co$_2$-emissions

Rising, J., Dietz, S., Dumas, M., et al. (2022). *What will climate change cost the UK? Risks, impacts and mitigation for the net-zero transition*. Grantham Research Institute on Climate Change and the Environment.

Sachs, J. D., & Ki-moon, B. (2015). *The age of sustainable development*. Columbia University Press.

Sakschewski, B., Cesar, L., Rockström, J., & Kitzmann, N. (2025). *Planetary health check 2025: A scientific assessment of the state of the planet.* https://doi.org/10.48485/pik.2025.017

Scoones, I., Stirling, A., Abrol, D., Atela, J., Charli-Joseph, L., Eakin, H., Ely, A., Olsson, P., Pereira, L., Priya, R., van Zwanenberg, P., & Yang, L. (2020). Transformations to sustainability: Combining structural, systemic and enabling approaches. *Current Opinion in Environmental Sustainability*, *42*, 65–75. https://doi.org/10.1016/j.cosust.2019.12.004

Steffen, W., Richardson, K., Johan Rockström, Cornell, S. E., Fetzer, I., Bennett, E. M., Biggs, R., Carpenter, S. R., Wim, d. V., Wit, C. A. d., Folke, C., Gerten, D., Heinke, J., Mace, G. M., Persson, L. M., Ramanathan, V., Reyers, B., & Sörlin, S. (2015). Planetary boundaries: Guiding human development on a changing planet. *Science*, *347*(6223), 1259855. https://doi.org/10.1126/science.1259855

Stockholm Resilience Centre. (2025). *Planetary Boundaries.* Planetary Boundaries. Retrieved January 29, 2026, from https://www.stockholmresilience.org/research/planetary-boundaries.html

UN General Assembly (2015). Transforming Our World: the 2030 Agenda for Sustainable Development. A/RES/70/1 United Nations.

UNEP. (2011). *Towards a green economy: Pathways to sustainable development and poverty eradication.* https://file:///C:/Users/Meri1/OneDrive%20-%20Middlesex%20University/Documents/Papers%20etc/Governing%20in%20the%20time%20of%20crisis/Literature/Sustainable%20development/UNEP%202011%20green_economyreport_final_dec2011.pdf

UNEP & CBD (2022) Decision Adopted by The Conference of the Parties to the Convention on Biological Diversity: Kunming-Montreal Global Biodiversity Framework. CBD/COP/DEC/15/4 https://www.cbd.int/doc/decisions/cop-15/cop-15-dec-04-en.pdf

United Nations Department of Economic and Social Affairs. (2025). *The sustainable development goals report 202.* https://unstats.un.org/sdgs/report/2025/The-Sustainable-Development-Goals-Report-2025.pdf

Vanclay, F. (2003). International principles for social impact assessment. *Impact Assessment and Project Appraisal*, *21*(1), 5–12. https://doi.org/10.3152/147154603781766491

Voulvoulis, N., Giakoumis, T., Hunt, C., Kioupi, V., Petrou, N., Souliotis, I., Vaghela, C., & Wan Rosely, W. (2022). Systems thinking as a paradigm shift for sustainability transformation. *Global Environmental Change*, *75*, 102544. https://doi.org/10.1016/j.gloenvcha.2022.102544

WCED. (1987). *Our common future: Report of the UN World Commission on Environment and Development.* Oxford University Press.

WEF. (2021). *Stakeholder capitalism: A brief history of GDP - and what could come next.* World Economic Forum. Retrieved December 4, 2025, from https://www.weforum.org/stories/2021/12/stakeholder-capitalism-episode-1-a-brief-history-of-gdp/

Whitmarsh, L., Player, L., Jiongco, A., James, M., Williams, M., Marks, E., & Kennedy-Williams, P. (2022). Climate anxiety: What predicts it and how is it related to climate action? *Journal of Environmental Psychology*, *83*, 101866. https://doi.org/10.1016/j.jenvp.2022.101866

World Bank Group. (2025). *June 2025 update to global poverty lines.* World Bank Group. Retrieved December 4, 2025, from https://www.worldbank.org/en/news/factsheet/2025/06/05/june-2025-update-to-global-poverty-lines

Zinecker, A., Gass, P., Gerasimchuk, I., Jain, P., Moerenhout, T., Oharenko A, Suharsono A, & Beaton, C. (2018). *Real people, real change strategies for just energy transitions.* ISIID. https://www.iisd.org/system/files/publications/real-people-change-strategies-just-energy-transitions.pdf

4 Governance and power

Learning objectives

- How to interpret governance terminology?
- What is relational power and what are its implications for good governance?
- How to distinguish between top-down, bottom-up and network governance and how can this support the analysis of governance and implementation gaps?
- What is meant by the constructivist critique of governance?
- What is reflexivity in governance and how does it support governance for transformational sustainability?
- What key theories are there for analysing global governance and what are their respective uses for identifying good governance for sustainable development at the global scale?

4.1 Introduction

This chapter builds on the definition of sustainable development as a *fundamental societal transformation towards equitable and regenerative relations among people and the planet, addressing the core drivers of unsustainability*. As discussed in Chapter 3, this implies a shift in the growth-oriented development rationale (Jackson, 2025) towards one that observes the Earth system boundaries (Folke et al., 2021) and guarantees everyone a sufficient standard for critical needs (Raworth, 2017). This sustainability transformation can be realised by centring the notion of care as an indicator of desirable economic activity and engaging with a combination of so-called green economy and degrowth solutions. But an evolution away from consumerist values and growth centrism is a necessary pre-condition. Governance needs to play a key role in facilitating this transformation.

Chapter 1 first defined governance as 'goal-oriented action regarding social coordination in the public domain' (Bevir, 2011) and subsequently set out a more precise deleniation of governance as *the institutions and practices of decision-making and intervention regarding issues of public interest, that is undertaken by formal and informal governance actors, often situated in*

DOI: 10.4324/9781003179207-5

networked relations (Morrison et al., 2019; Sørensen, 2002). This chapter unpacks this definition and introduces a range of theories and approaches that can be used to analyse governance and its ability to support sustainable development.

Sustainable development as societal transformation evokes a range of conflicts of interest (Scoones et al., 2020). As suggested in Chapters 2 and 3, these societal trade-offs result from the difficulty of reconciling the prevalent economic growth imperative with the need to observe the limits of the planetary system. The so-called yellow vest protests that brought France to a standstill in 2018 serve as an example highlighting the subsequent importance of legitimacy and justice in governance for sustainable development (Oxfam, 2023). The protests, initially by transport industry workers, were sparked in response to a 'green tax' on petrol intended to reflect the environmental cost of carbon emissions but were perceived as grossly unfair in the context of increasingly precarious employment conditions in the transport sector and a rising cost of living (Driscoll, 2023).

Governing for sustainable development therefore requires addressing difficult moral dilemmas in a highly politicised context. With many key democratic institutions such as free elections and press freedom in decline globally, the legitimacy and transparency of governance goals, processes and outcomes are crucial. But growing inequality feeds societal discontent and opens opportunities for populist discourses and leaders to gain ground. If governance is to support a just transformation for sustainability, the integration of multiple perspectives and knowledges into decision-making and a fair and transparent reconciliation of conflicts of interest are key requirements.

Legitimacy (Sell, 2018) stems not only from formal governance actors and processes, but from the fairness and public acceptance of governance aims, processes and outputs. Therefore, the workings of power – the interrelations between governance actors (including policy recipients) are decisive for good governance. But the substance, the issues that governance focuses on, is also significant. Here, we start from the premise that good governance needs to also yield outcomes that are consistent with the principles of sustainable development (Weiss & Wilkinson, 2018).

4.2 Governance terminology

Governance literature, including the above definition of governance, is laden with jargon. This section defines some of its key terms to facilitate the reading of subsequent sections and chapters. Firstly, 'governance institutions' refer to tangible actors such as individuals, groups and organisations, as well as less tangible but no less influential things like legislation, regulations, guidelines, standards and rules of operation and collaboration. 'The practices of decision-making' in turn refer to what is perhaps most instinctively understood as governance and sometimes also referred to as the 'design phase' of governance where decisions regarding the approach and instruments of governance are made (Morrison et al., 2019). There is a range of ways of gathering evidence and arriving at decisions regarding governance 'outputs' encompassing expert-led or collaborative negotiation and consultation (Cairney, 2020).

'Governance outputs' (also termed 'governance instruments') in turn refer to the strategies, policies, plans, standards, permits, subsidies, taxes, projects and programmes to be 'implemented' in practice to achieve the desired behavioural change. These are distinct from 'governance outcomes' meaning the impact resulting from governance outputs – the resulting changes in behaviour such as the adoption of new technologies and subsequent decrease in emissions, for example. Any governance output that is actually applied in practice will evolve in implementation, as the involved actors (including those intended to 'comply' with the desired behaviour changes, often termed 'governance recipients') interpret and appropriate its aims (Juntti & Potter, 2002).

It is helpful to think of governance in terms of a cyclical process encompassing agenda-setting, decision-making, implementation of outputs and monitoring and feedback on the outcomes of governance (Cairney, 2020). While 'issues of public interest' entail a broad range of concerns from trade and human rights to health and environmental issues, many of them are now brought together under the umbrella term of sustainable development (see Chapter 3). The issues that governance addresses are also often termed 'public goods' such as clean air, a healthy environment and peace and political stability (Brousseau et al., 2012a).

'Formal and informal governance actors' is an important distinction between stakeholders who have a legislative mandate and those who may be equally powerful but do not. Formal actors are often directly affiliated to or instated by the state (Bevir, 2011), such as governments and their departments and regional and local agencies. Informal actors are involved because they wish to advocate for specific interests or solutions or by invitation if the formal actors believe they have information, influence or resources that will benefit the outcomes of the governance process. The latter group encompasses a broad and diverse range of non-state actors such as local or international interest organisations, businesses and, in some cases, even individual people.

Finally, a distinction is often drawn between governance and *global governance*. In literature generally, global governance refers to concerted action on the part of national governments and informal and international scale actors on issues that necessitate inter-jurisdictional collaboration (Bevir, 2011). Here, global governance refers to the collaborations at a global scale, between actors such as nation states, international governmental and non-governmental organisations and private entities such as corporations, focusing on what can be termed ‘global public goods’, issues of significance at the planetary (or Earth system) scale, such as climate change and food security (Brousseau et al., 2012).

4.3 Power in governance

Governance, referring to the institutions and practices of decision-making and intervention regarding issues of public interest, plays a key role in enabling a transformation towards a fairer, more sustainable and more stable society. Interrogating the notion of power is helpful for understanding how societal change can be achieved and sustained. Power can be understood in different ways but ultimately, what many writers perceive at its core is the ability of one social actor to persuade others to modify their behaviour in a certain way (Castells, 2004).

Stephen Lukes offers a foundational framework for understanding power, challenging common assumptions about what it means to be ‘in power’ (Lukes, 2017). Power is conventionally understood as emerging in situations that present a conflict of interests, person A gets person B to do something B wouldn’t otherwise do. But Lukes highlights the notion of ‘non-decision-making’ as a form of power also. Non-decision-making refers to the exclusion of select issues from the decision-making agenda altogether despite their significance to the public interest. Entrenched values and institutional practices of decision-making within administrations often influence both what is seen as ‘usable knowledge’ that should inform governance decision-making and what issues are considered ‘governable’ (Haas, 2004). These practices and perceptions therefore play a role in the realisation of power.

Governance actors – formal or informal – are not neutral but often exhibit powerful policy preferences. For example, in Spain, the agricultural ministry persistently limited the focus of governance efforts to water *quantity* rather than *quality*, significantly slowing down progress in implementing the EU Nitrates Directive, which addressed the *quality* of water (Izcara Palacios, 1998). Indeed, environmental governance actors and scientific evidence of environmental issues often struggle to gain legitimacy in agricultural policy arenas (Juntti & Potter, 2002). This may result from internal, institutional bias as in the above example. But bias may also be introduced through lobbying, for example where farmers’ unions have traditionally held close ties to formal decision-makers (Juntti & Potter, 2002). Non-decision-making and lobbying can corrode the legitimacy of power, where legitimate power aligns with shared values and is acceptable to those being governed (Parsons, 1986).

In line with this, Lukes further refines power as a collective capacity, not control or coercion by a powerful actor. Hannah Arendt argues that the need to use violence, or coercion, in fact signals a lack of legitimacy and therefore of power, as ‘true power’ arises from collective agreement and consent (Arendt, 1986). In line with this, policies that work without the need for punitive measures are powerful and signify good governance, and those requiring constant enforcement lack legitimacy and therefore power. To capture this, Lukes proposes a three-dimensional view of power, encompassing systemic and institutional dynamics (Lukes, 2017). Power manifests through organisational structures, cultures, shared norms, and institutional practices that shape decision-making and outcomes, without visible conflict. Institutional structures and practices determine, for example, the way evidence is gathered for decision-making, who is consulted and how (Haas, 2004). Cultural norms, in turn, determine what is seen as 'common interest', as with the agricultural example above. Cultural norms may stem from the professional background of governance actors – whether they have expertise in agronomy, environmental science or public administration, which also influences what kind of evidence they are used to interpreting and inherently trust.

Lukes also draws attention to influence and authority and the manner in which they frame the issue at hand and shape perceived interests, influencing both formal and informal governance actors, as well as the so-called recipients of governance, and subsequently the perceived legitimacy of governance interventions. Media, politicians and advocacy groups exercise this sort of framing power, a dynamic that is central to understanding populism and 'false consensus' (Morrison et al., 2019). Ultimately, Lukes shows that power is not just about winning in times of conflict, but also about what gets excluded, how interests are shaped, and how legitimacy is constructed (Lukes, 2017).

Writers such as John Allen (Allen, 2011) take this approach further to argue that power is vested in interaction and that it is the networked relations of actors – human and non-human – that are decisive to the operation of power. According to this 'relational perspective', a governance actor or instrument is 'powerful' when it is 'made present' in the everyday lives of those whose actions it intends to influence. Returning to the agricultural example, this would mean that the EU Nitrates Directive is powerful only when its aims are enacted in the everyday lives of farmers managing agricultural land. And that this will only happen once the whole network of actors, from the EU Commission in Brussels through to the Agricultural Ministry in Spain and the farmers, their machinery and even the nitrates in the field collaborate to ensure that no nitrates make their way into water courses polluting drinking water and causing eutrophication (Chapter 6).

This notion of power as a network quality highlights the connections between actors as crucial for the operation of power. These connections consist of flows of information and practices of collaboration. It highlights the power of technologies such as electronic communication methods that make it possible for information to be produced and shared instantaneously over distance. This way of understanding power reveals a range of powerful influences that shape our everyday lives. But staying with the above example of nitrate pollution from agricultural run-off, this could refer for example to governance interventions that rely on soil quality analysis, where farmers are provided chemistry kits to assess the nitrate content of the soil that they cultivate (Bellosta-Diest et al., 2022). Being able to perform soil analysis in the field makes nitrates 'present' in the everyday management of farmland and empowers prudent application of nitrate fertiliser and the use of soil management techniques, which will minimise nitrate leaching and run-off. Here, communication and technology further engage cultural norms, according to which attention to soil quality is seen as 'good farming practice'. The use of such technology therefore is likely to be an acceptable practice to farmers, yielding the intervention legitimacy. This contrasts with a command-and-control approach based on the notion of power as authority, which would seek to control farmers' actions by posing a threat of punitive measures if a sanctioned amount of fertiliser was exceeded for example. Many argue that the outcomes of networked rather than hierarchical governance relations are more durable and therefore more powerful.

In order for governance interventions to work in practice, we need to consider the real-world decision-making of the key actors and for this, we need to recognise the mediating entities of power, such as technologies, information, guidance, regulations, laws, funding programmes, administrative practices, practices of decision-making, customs among policy makers and recipients - culture and counter-culture (Latour, 2004). The mediating entities of power are often non-human, and non-human entities such as technology and natural resources play a significant role in empowering humans to act in certain ways (Latour, 2004).

In terms of lessons for good governance, this understanding of power as relational and as realised through networked constellations among heterogeneous actors highlights the importance of explicitly acknowledging that effective governance involves a multiplicity of perspectives and, in order to succeed, needs to engage with and build relations between all the relevant actors (formal and informal, human and non-human). It draws attention to the power of inanimate and non-human subjects in governance, such as technology and natural entities, and the influence of cultural norms and practices often detectable in the shape of different discourses (ways of speaking about governance issues).

4.4 Governance approaches and power

This section considers the most prominent governance theorisations that operationalise and align with the different understandings of power discussed above.

4.4.1 Top-down and bottom-up approaches to governance

Governance can be understood in terms of a dichotomy of hierarchical top-down and bottom-up community-led action (Ham & Hill, 1993; Sabatier, 1986). The top-down approach regards power as concentrated at the top of the policy hierarchy, with consultations limited to expert stakeholders (such as scientific experts) whose role is to provide advice to minimise any potential implementation defect. Clarity of governance goals and rules, and an efficient implementation structure are seen as central to success (Ham & Hill, 1993). This approach subscribes to a territorial perception of power, referring to the territorially bound authority of national executives and stemming from the notion of strong state sovereignty (Allen, 2011).

The bottom-up approach (Rhodes, 2007; Sabatier, 1986) on the other hand, starts with the operational level of governance, in what can be termed its site of implementation. The nature of the governance issue at hand and the strategies of the relevant stakeholders in the pursuit of their own objectives inform bottom-up governance (Sabatier, 1986). In fact, the bottom-up approach can be said to focus on understanding and working with precisely the factors that the top-down approach aspires to minimise.

In rendering the 'operative level' as the focus of attention, the bottom-up approach aligns better with the relational understanding of power. Both recognise that behavioural decisions are informed by a range of interests and practicalities. While the bottom-up approach focuses on governance recipients and their realities and interests, the notion of relational power fine-tunes and extends this focus to the role of hybrid actors and relations that tie them to the site of governance outcomes. Returning to the above example of the attempts to govern nitrates and water quality in agriculture (Izcara Palacios, 1998), the relational perspective emphasises the importance of technologies such as soil nitrate testing kits but also knowledge exchange, scope for negotiation, and monitoring activities for the success of any governance intervention.

According to the bottom-up relational understanding of power, the design and implementation of a governance intervention need to fit with the rationale and everyday realities of governance recipients. This necessitates the opening up of the design and implementation phases to different types of knowledge – the practical everyday knowledge of 'living the field' and the expert scientific knowledge that provides a window to the behaviour of nitrates in different soil types (Kaljonen, 2006). Only by taking these perspectives seriously can we arrive at governance interventions that will persuade farmers and nitrates to collaborate for the benefit of crop production, without nitrates escaping to ruin water courses (Juntti, 2012).

This realisation has led to the emergence of a large literature on 'networked' or 'polycentric' governance (Morrison et al., 2019; Ostrom et al., 1999) and the idea of 'reflexivity' – the ability to draw from a broad range of relevant perspectives and to engage with complexity as central to good governance (Voß & Kemp, 2006). The shift from a hierarchical and vertical top-down approach to a horizontal network approach appreciates power as dispersed through the governance process and assumes that governance interventions evolve in implementation thanks to a range of influential forces (Morrison et al., 2019).

4.4.2 Network governance and reflexivity

'Network governance' refers to diverse stakeholders collaborating to set and achieve governance goals (Bulkeley, H., 2012; Morrison et al., 2019; Sørensen, 2002). In network governance, spatial and issue boundaries (for example, a water catchment comprising the land area from which rainwater drains into a surface water body) rather than administrative units, such as a nation state, region, or municipality, define the scope of these collaborations and governance interventions (Bulkeley, 2012). These governance networks collaborate in horizontal rather than vertical and hierarchical governance arrangements. Network governance is an increasingly prominent approach in environmental governance where collaboration between formal and informal actors across different sectors of expertise and experience (such as the environmental and agricultural sectors, for example) is increasingly common (Bulkeley & Mol, 2003; Evans & Thomas, 2024).

There are various theorisations concerning why and how governance networks come about and what constitutes good governance in this context. Formal and informal linkages between governance actors are constructed due to shared interests (Rhodes, 2007). Governance networks are durable because participants gain something from their membership – for example, increased ability to meet a shared goal, achieve legislative compliance, or a chance to influence up-coming policy. It is evident that shared values, clear norms, and trust are important for the success of any governance network. This makes sense from the perspective of relational power also: in order for the governance network to operate successfully, there needs to be a shared understanding of the goals of governance, and these need to be stabilised into consistent guidance and practices.

But Rhodes (Rhodes, 2007) suggests that even in networked governance relations, there is a core executive, which coordinates and integrates central government policies, arbitrates conflicts, and delimits participation in the governance process. This would mean that institutional biases may continue to inform which issues will be included or excluded from the governance agenda and shape the behaviour of the networked governance actors through the set rules and norms, privileging certain perspectives and types of evidence (Haas, 2004). One example of this is the imposition of performance metrics which fit some types of governance solutions better than others and therefore hinder the integration of multiple perspectives and needs (Taylor, 2007).

For example, the requirement to demonstrate value for money in governance may translate into the need to rely on governance instruments such as blanket emission standards, which do not take into consideration contextual variation in vulnerability or individual governance stakeholders' circumstances, but are easy to implement and quick to monitor. This is not necessarily a bad thing from the point of view of accountability and transparency, which tend to be the weaker aspects of network governance

(Rhodes, 2007). But it means that networked governance arrangements may not be as collaborative in real life as they seem on paper (Taylor, 2007).

At times, a number of formal government agencies may be involved in governance networks, which gives rise to 'polycentricity' (Morrison et al., 2019). This too brings both benefits and challenges. While polycentricity embraces the complexity that characterises sustainable development issues by including multiple different actors and perspectives, governance networks can also exacerbate conflicts between parties. In the absence of a clear executive authority, this can result in decision-making 'gridlock' or implementation deficit (Rhodes, 2007). Morrison et al. (2019) suggest that it is helpful to think of power in polycentric governance as consisting of distinguishable spheres, which draw together both formal and informal governance actors. The three spheres pertain to influencing the normative and issue context (framing power), making decisions about the governance outputs (design power), and the manner in which they are implemented (pragmatic power). These dimensions of power can be seen to align with the different stages of the governance process (Cairney, 2020).

Finally, Rhodes (Rhodes, 2007) draws attention to a lack of evidence of how governance networks integrate change. In fact, they may be fairly change resistant, thus falling short of the requirement for reflexivity – the ability to respond to changing circumstances and feedback loops pointing, for example, to knowledge gaps or non-decision-making. Possible change forces may be exogenous, such as economic, ideological, knowledge and institutional factors, or result from persistent dilemmas or ideological change that arise in the context of everyday governance activities (Rhodes, 2007). Path dependency refers to rigidity resulting from past choices, such as established governance instruments and targets and associated sunk costs. Here, framing power and the role of media and public discourse are of particular importance as a potential change force.

4.4.3 The constructivist approach to understanding governance

In order to understand the evolution of governance and embrace the relational nature of power, we need to understand governance through a constructivist lens, which brings to the foreground the influence of ideas and norms and the way in which issues are framed within governance. As discussed above, problem framing is a form of power and often favours one stakeholder perspective over another in recognising certain issues and challenges (Morrison et al., 2019). Rather than a model of governance, constructivism is a methodological approach, based on the ontological assumption that reality is socially constructed rather than consisting of immutable structures (Park, 2018). Focusing on the agency of these social constructions in governance, it aligns with the relational approach to power in that it recognises that there is more to governance than the power reputation of formal governance actors. Lukes's three-dimensional view of power reflects a constructivist approach.

Constructivism has enabled scholars to understand the emergence of powerful ideas such as 'sovereignty', 'justice' and now, 'sustainability', and to interrogate their evolution and influence within global governance. By drawing more attention to the power of ideas, and the 'norms' that they often evolve into, constructivist research has enabled us, for example, to better articulate the position of developing countries in the UNFCCC and identify the tactics which so-called developed countries use to resist the calls for reparation justice and climate restitution payments (UNFCCC, 2022).

The constructivist approach has also enabled a better understanding of the role of knowledge in governance. For example, understanding how governance actors define 'usable knowledge' has highlighted the need for scientific and other types of knowledge in decision-making. And the notion of 'epistemic communities' refers to groups or networks of like-minded knowledge producers (scientists but increasingly also those in possession of experiential or 'lay' knowledge) that wield power in the knowledge-governance interphase (Haas, 2004). In the context of the crisis of knowledge, the constructivist approach alerts us to the need to stay vigilant to both the dangers posed by too narrow a knowledge basis on one hand and populist manipulation of knowledge on the other (Sze, 2020).

4.5 Global governance for sustainable development

Sustainable development is a global cause that requires global collaboration. While the world is becoming politically fragmented, we remain interdependent through global commodity networks, such as the international food system, and need to coordinate actions pertaining to global public goods such as climate and health (Brousseau et al., 2012b). The UN resolution setting out the Sustainable Development Goals (UN General Assembly 2015) is one such example, and a global governance output. Further examples are the UN Framework Convention on Combatting Climate Change (UNFCCC) and the UN Convention on Biological Diversity (CBD).

Both the UNFCCC and teh CBD emerged in the wake of the Earth Summit in Rio de Janeiro, Brazil in 1992. The UN and its Environment Programme (UNEP) remain the main forum for global governance for sustainable development. But the European Union (EU) constitutes a significant regional intergovernmental organisation furthering environmental sustainability and human rights globally, and the Organisation for Economic Cooperation and Development (OECD) is another (O'Neill, 2022). Multinational Corporations (MNCs) and Non-Governmental Organisations (NGOs) are also active on the global arena, as are international issue coalitions and epistemic communities (Haas, 2004). The following sections outline some theorisations of global governance that help us understand and analyse the activities of these and other organisations on the global arena.

4.5.1 International regimes in understanding interstate collaboration at the global scale

The international regimes theory stages nation states and international organisations as key actors in global governance and draws attention to how multilateral agreements emerge and influence the actions of their 'Parties', the states that have ratified them. The regimes usually produce so-called international or 'soft' laws consisting of formal and informal rules, procedures and norms that pertain to a specific issue area (Weiss & Wilkinson, 2018). One such example is the UN Framework Convention on Combatting Climate Change (UNFCCC), explored in Box 4.1. According to the international regimes theory, states are motivated to form multilateral agreements by the benefits vested in interstate cooperation, such as better access to information and improved ability to negotiate, share knowledge and know-how as well as better guarantees that all involved states comply with commonly agreed targets and rules (Sell, 2018).

The international regimes theory draws attention to how intergovernmental consortia approach and establish principles, norms, rules and decision-making procedures. For multilateral agreements to function and achieve set goals, they need to formulate roles and rights and establish a degree of organisation surrounding issues of enforcement or compliance (Bohman, 1999). Regime theorists have grappled particularly with the challenges of democracy and legitimacy experienced by these regimes. In the absence of a 'global public' central tenets of good governance such as accountability and legitimacy are difficult to conceptualise let alone achieve at a global scale. Equal access to political influence is difficult to guarantee, and some have even suggested that this may be a redundant ideal (Bohman, 1999). The democratic and legitimacy gap in global governance has meant that issues, decisions and their underpinning rationale often depart from the perceptions and experience of those affected. This corrodes the ability of global treaties to achieve the desired outcomes. The lack of legitimacy has inspired broadening involvement of non-state actors in global governance from the 1990s onwards (Sell, 2018).

Box 4.1: The UN Framework Convention on Combatting Climate Change (UNFCCC) – an example of an international regime

Initiated at the UN 1992 conference at Rio and in force since 1994, the Convention is ratified by 197 of the UN member states. Under the Convention, governments agree to:

1. gather and share information on greenhouse gas emissions, national policies and best practices

2. launch national strategies for addressing greenhouse gas emissions and adapting to expected impacts, including the provision of financial and technological support to developing countries
3. cooperate in preparing for adaptation to the impacts of climate change.

Initially, the convention encouraged industrialised countries to stabilise climate change causing emissions and was based on the principle that national targets are to be met through national measures, rather than centrally prescribed actions. The parties to the convention, the 197 nation states that have signed up to it, report on progress towards the set targets and their implementation to the Conference of Parties (COP), and thus, the convention is enforced through the 'naming and shaming' of laggards, who do not comply with the commitments.

While based on what can only be termed a weak enforcement mechanism, the Convention failed to achieve cuts in emissions in its first two decades. However, in 2015, at the Paris COP, mainly thanks to growing pressure from scientific advisers in the International Panel on Climate Change (IPCC) and a change in public opinion, tangible progress was finally made. By far the majority of the Parties (193 with the notable exception of the US under the Donald Trump administration) signed up to the Paris Accord (UNFCCC, 2022), pledging to set necessary emission reduction targets to limit global warming to 2°C, with an aspiration for 1.5°C. Within the subsequent 5 years, most large emitters, such as EU countries, the UK, the US (who joined as soon as President Trump's first term was over), Australia and finally even China established legal guarantees to become carbon neutral (limit emissions to no more than what is absorbed in CO_2 equivalent, though other activities such as sequestration and existing green cover) by the second half of the century. In 2021, COP26 went further to outline the specific measures that all parties will adhere to, such as stopping deforestation, cutting methane emissions by 30% by 2030, increasing the production of zero-emission vehicles, moving away from coal to renewables-based economies, and directing private investment into the renewables sector (UNFCCC COP26, 2021).

The Convention has thus forged out some key principles for climate mitigation and adaptation. It has also established a normative framework delegating responsibility for these activities in accordance with the different Parties' levels of development and responsibility for climate emissions in the first place. Since Paris, there has been significant onus on developed countries (OECD members) to provide funding and share clean technology with developing countries. Restitution payments (compensation for damage and loss that is ongoing and that has already been suffered) were also on the agenda in Glasgow, and a decision in principle was reached in Cairo at the COP27 to establish them, but no tangible funding has as yet materialised. Rich countries remain reluctant to commit money, it seems, and questions have been raised over whether some of the countries listed as developing (Originally Non-Annex I Parties) in the convention, such as China, should still be classed as such.

For more information on the Convention, visit: https://unfccc.int/process-and-meetings

4.5.2 Multilevel governance

A critical, relational understanding of power refutes the idea that global governance is based on a neutral hierarchy of national, regional and local level actors and that power is concentrated at the top, the global scale. After all, interstate collaboration at the global scale is significantly influenced by the internal affairs of each state, including political leanings, economic agendas and degree of international influence (Weiss & Wilkinson, 2018). Similarly, the outputs and outcomes of global agreements are shaped by national governance apparatuses, their structures and cultures, in implementation.

The multilevel governance approach directs attention to the various scales of governance beyond the state level, and the types of power constellations that emerge among scalar actors, either across sectoral jurisdictions or pertaining to just one jurisdiction such as health or the environment. Multilevel governance recognises the dispersion of authority to supranational, regional and local levels (Hooghe & Marks, 2004). Multilevel governance can be informed by the constructivist lens, to acknowledge that the hegemony of nation states and their administrative bodies is tempered by a range of influential institutions, norms and agendas. This approach emphasises the role of private-public partnerships and non-governmental actors at national and international levels. Polycentric and multi-perspective governance and overlapping or competing jurisdictions are all theoretical models that can help bridge the gap between the global and the local scales in governance. Hooghe and Marks (2004) outline two alternative models of multilevel governance:

1. One relevant jurisdiction at any territorial scale, where the basis of this approach lies in the theory of federalism. Here, general-purpose jurisdictions, for example, a regional government, exercise a wide range of functions and provide all-round care for its citizens. The European Union (EU) is an example.
2. Specialised jurisdictions that provide a particular service pertaining to a specific issue or scale. Multilateral conventions, such as the UN Convention on Combatting Desertification (UNCCD), are examples.

As an example of Type 1 multilevel governance, the EU (Eising, 2015) demonstrates the power of central institutions, such as the EU Commission and the Parliament, in supra-national decision-making. Sub-national actors, for example, city administrations or sectoral interest groups, can also influence EU governance decision-making, bypassing national governments, as it were. Indeed, 'lobbying' is particularly prominent in Brussels, where interest groups frequently target the European Commission, which has an influential role in proposing and preparing new legislation (Lundy, 2017). EU institutions, such as the European Parliament, the EU Commission, and the European Court of Justice, are independent from state influence. The Type 1 multilevel governance model suggests that the EU's scalar governance arrangements should be viewed as an interlinking network rather than nested. For example, the European Council, within which the Member State governments wield power, must collaborate with the Parliament of directly elected members of the European Parliament (MEPs) through a process termed co-decision (Farrell & Héritier, 2004).

As for Type 2 multilevel governance, the issue-based multilateral conventions, such as the UNCCD, serve as examples. These jurisdictions are not clear-cut in terms of boundaries and may overlap, forming a much more complex map than Type 1 systems. These task-specific jurisdictions operate at a number of scales, and governance actors include a broad range of formal and informal governance actors. This mode of organisation is flexible and dynamic, meaning that governance organisations may not be long-lived but emerge according to demand on a project basis. Public-private partnerships are common, as with the World Bank and International Monetary Fund, and many multilateral agreements such as the UNCCD, where private money is increasingly sought to address the provision of public goods, such as afforestation projects or sustainable cultivation methods (UNCCD, 2025).

In this manner, the multilevel governance approach highlights the vertical relations of power in global governance and has been adopted particularly enthusiastically to better conceptualise and understand what is happening within the EU – a fairly unique regional governance entity. But the principles and analytical units offered by the approach are also useful for understanding how what in a hierarchical model appears as 'lower' governance scales, such as city administrations, can at times drive change at the national or even international scale, for example through pioneering innovations such as incentives for energy efficiency measures that inform global scale discourse (Kuokkanen & Yazar, 2018). The agency of city administrations may be realised through transnational networks that bypass national governments, such as the Local Governments for Sustainability (ICLEI) network (ICLEI, 2025).

The appreciation of complexity in how power operates in global governance and links to the local scale and vice versa has also inspired some to suggest that in fact governance and economic domains are strongly intertwined, because political interaction is one of the principal means through which the economic structures of the market are established and transformed, and economic interests in turn shape political interactions (Weiss & Wilkinson, 2018). Many also draw on critical theory such as feminist, 'Gramscian' and environmental schools of thought to develop a more fine-tuned understanding of the workings of power associated with the expanding but uneven agency wielded by networks of sub-national actors as well as global institutions such as the WTO, IMF, and the World Bank, and international financial institutions such as regional development banks, credit rating agencies, insurance brokers, and corporations in global governance (Watson, 2014).

4.5.3 Architectures of Earth system governance

The interdisciplinary 'Earth system governance' framework draws on all of the above approaches and supports the understanding and assessment of the institutions, interactions and policy responses pertaining to governing the 'Earth system' (Folke et al., 2021). The main aim of this approach is to assess the performance of global environmental governance from the perspective of its problem-solving capacity, legitimacy, accountability and equity (Biermann & Kim, 2020). It is an ambitious model and perhaps the first of its kind in that it stipulates the need for complex systems thinking, such as observing the Earth system boundaries, in governance. It acknowledges therefore that most of the so-called environmental issues really pertain to the whole Earth system which incorporates the biosphere and society, as well as the non-living parts of the Earth system (such as the climate and mineral resources) (Folke et al., 2021).

Because this approach recognises and articulates complexity and interdependency as perhaps the two most prominent characteristics of the Earth system, it draws our attention to governance challenges that remain neglected or are only incrementally recognised in other approaches. An example of these is 'governance fragmentation', resulting from uneven regulatory contexts of nation states. This enables 'problem shifting' between international regimes, which underpins inefficiencies and 'leakage' of issues, such as carbon emissions or deforestation, that persist despite global efforts to govern them (Biermann & Kim, 2020). An example of issue or problem leakage is the tightening of forest conservation measures in Vietnam in the 1980s, which led to the 'leakage' of deforestation to its neighbouring countries such as Cambodia, where the sector is less stringently regulated (Bastos Lima et al., 2019). As a result, deforestation continues despite an increasing number of countries signing up to the global forest conservation initiative REDD+.

Earth system governance also advocates for a network-based approach to governance. Literature suggests that institutional interplay and its management through collaboration is a crucial aspect of global environmental governance and that this needs to happen across scales (from local to regional and international) and administrative and issue boundaries (e.g. across climate change and desertification policies and regimes). As the example of the

ICLEI network of cities for climate mitigation suggests, sub-national and transnational scales are often the two most important focal points when trying to understand transitions to sustainable societies (Kuokkanen & Yazar, 2018).

Notions such as institutional fragmentation and issue leakage can alert us to governance inefficiencies where seemingly powerful governance solutions fail to reconcile crucial societal trade-offs. This is often due to the ability of certain stakeholders to engineer governance solutions that serve their sustainability reputations while avoiding the cost of actually solving the problem. In the above example from Vietnam, the reputation of the Vietnamese forestry industry benefited from the tight environmental regulation but the 'leaking' of deforestation to neighbouring countries continued to guarantee a source of cheap timber for its purposes (Bastos Lima et al., 2019). The end result was a failure to resolve the problem of deforestation at a global scale. Better management of the interplay of the Vietnamese and Cambodian forest governance approaches and institutions would likely have helped, but there was no political will or indeed forum for this. Therefore, the issue at hand – loss of biodiversity and carbon sink capacity – 'escaped' the REDD+ framework, which was in this case rendered powerless.

4.6 Power, sustainable development, and good governance

As Weiss and Wilkinson (Weiss & Wilkinson, 2018) suggest, good governance is about more than just the right kinds of constellations for decision-making and implementation. Also, the aims of governance matter. Good governance needs to advance sustainable development in a fair, legitimate and transparent manner. The notions of distributive, procedural and recognition justice discussed in Chapter 3 are helpful criteria, and good governance needs to observe the Earth system boundaries (Folke et al., 2021). But beyond this, good governance can be described as governance that works – governance outputs that achieve the intended fair, legitimate and sustainable outcomes. A critical understanding of power and governance arrangements and processes is central to determining good governance.

Governance efforts pertaining to sustainable development are challenged by the 21st-century crises pertaining to environmental, social and knowledge-related complexities (see Chapter 2). While socio-economic inequality is accelerating, inequality is also a concern across the drivers and impacts of environmental problems as well as ability to withstand them (to adapt and to remain resilient). Moreover, environmental issues necessitate reliance on complex and often interdisciplinary scientific knowledge, evoke conflicts of interest and impact stakeholder groups in highly differential ways (Turnpenny et al., 2009). This makes them ripe for discord and complicates decision-making regarding governance options. The ability of science and experts to 'lead the way' in governance for sustainable development is increasingly challenged. In the 21st-century governance context, political and value-based statements are perceived as legitimate arguments, but at the same time, we need to avoid the dangers of populism (Sze, 2020), which is likely to compromise measured, plural and fair governance interventions and outcomes.

Whether we like it or not, there is a widening range of actors that have a say. Good governance will need to embrace the plurality of values, perspectives and knowledges (without falling into the trap of populism – see Chapter 2). The notion of relational power helps us appreciate what this means in the context of governance. If we imagine that power is dispersed in a hybrid network of actors, and good governance equates to policies that achieve their stated aims, then it becomes evident that governance should focus on coordinating and facilitating the actions of those networks of people and things rather than on designing and enforcing behavioural models top-down. To keep everyone on board, expert claims concerning the state of the environment must be weighed in the context of conflicting interests and risk positions, and increasingly, partisan policy solutions promoted by different influential actors. There is also the preconceived institutional critique by non-governmental organisations and the demands and assertions by 'street level bureaucrats' that bring plurality and introduce different types of knowledge (scientific but also experiential) to decision-making rendering 'evidence-based policy' really quite political (Grin, 2006).

'Reflexive governance' (Voß & Kemp, 2006), sets out a framework for good governance that acknowledges the above complexity and embraces the relational nature of power. Reflexive governance recognises the need for value plurality and deliberation in decision-making, as well as the inclusion of different types of knowledge beyond the scientific perspective. Reflexivity implies not only cyclical development of governance interventions and solutions through trial and error but recognises that there are multiple possible premises of decision-making and that selecting one specific premise should be a conscious and informed decision, which governance actors should review regularly (Voß & Kemp, 2006). This refers to the different types of 'logic' or 'rational' thought that different governance actors rely on.

Returning to the example of the EU Nitrates Directive, the Spanish Ministry of Agriculture brought a sector-specific rationality to water governance, centring the focus on water availability in decision-making (Izcara Palacios, 1998). This initially excluded concerns for water quality from the governance agenda. Another perspective is that of 'formal rationality' where the principles of 'new managerialism' (Rhodes, 2007) come to the fore, bringing economic considerations and 'value for money' to the forefront of decision-making, emphasising the need for efficient governance mechanisms and measurable outcomes. Reflexivity suggests that while important, neither of these rationalities alone should dominate decision-making. Reflexivity involves a conscious effort to integrate a range of relevant perspectives and rationalities, from scientific to expert and 'lay' or experiential perspectives, including, where relevant, Indigenous ones which often frame environmental considerations as integral.

Therefore, while scientific evidence is crucial for the resolution of environmental issues, reflexive governance avoids narrow reliance on scientific evidence and techno-centric solutions, which provide the illusion of control. Reflexive governance principles suggest that instead of trying to eliminate uncertainty, ambivalence and interference from context-specific factors, governance needs to be able to accommodate them. In fact, complexity minimisation is associated with unpredictable governance outcomes, and instead, good governance requires setting aside specialised problem orientations (Beck, 2006). Reflexive governance emphasises learning, where governance actors and processes remain open to a range of different rationalities and confrontations between these and are focused on modulating ongoing developments rather than achieving often illusive control over them (Dedeurwaerdere et al., 2012).

To sum up, to accommodate a plurality of rationalities, knowledge and possible outcomes, reflexive governance promotes a deliberative approach in decision-making. Interactive deliberation rather than one-way communication to which policy consultation is often reduced allows for actively seeking out and exploring conflicts of interest, uncertainties, ambivalences, and control problems. Here, networked governance arrangements provide a good basis for reflexivity. In practice, reflexive governance focuses on problem analysis, goal definition, assessment and strategy implementation, engages with transdisciplinary research (that treats the perspective of lay/non-expert stakeholders such as communities as valid knowledge), implements robust ex-ante impact assessments, foresight exercises and makes room for deliberation in decision-making. Learning is maximised when the problem perceptions, interests, norms and action strategies of different actors are exposed to each other. This is because learning happens through interaction, conversation, conflict resolution and deliberation (Voß & Kemp, 2006). One way of supporting reflexivity in practice is provided by citizens' assemblies (Boswell et al., 2023), which invite members of the public to regular assemblies that inform governance. Citizens' assemblies can feed into the design and pragmatic phases of governance and play a role in framing governance issues. By convening groups of 'ordinary' or 'lay' participants from all walks of life, they can act as platforms of conversations and deliberations that initiate change in cultural norms and principles away from the current consumerist values (Mcdermott & Surminski, 2018). While citizens' assemblies are becoming more frequent in integrating plural perspectives into climate governance, their ability to influence formal governance actors remains debated (Boswell et al., 2023).

4.7 Conclusions

If we take sustainable development to mean *fundamental societal transformation towards equitable and regenerative relations among people and the planet, addressing the core drivers of unsustainability and altering their underpinning relations of power* (see Chapter 3), governance inevitably evokes a range of conflicts of interest. This means frequent struggles over what is accepted as governable and what counts as usable knowledge. To guarantee good governance that yields fair and sustainable outcomes, we need to pay attention to the workings of the tacit power of governance practices, institutions and non-human governance actors. The ability of governance actors to flexibly adjust to unexpected governance outcomes and innovatively reconcile plural values and perspectives is also paramount.

There are various different theoretical lenses that have been developed to understand and analyse governance. In order to fully appreciate the complexity and dynamic context that decision-makers are dealing with today (see Chapter 2), we need to interrogate the true nature of power in governance. The notion of power is central to governance, and the relational understanding of power (Allen, 2011; Latour, 2004) helps draw attention to the role of, not just human power struggles but of non-human entities in shaping our problem definitions and rendering governance solutions either powerful or not. Realising that power is in fact always enacted in interaction between hybrid actors points to networked governance arrangements that often bypass the conventional hierarchy of state-led governance constellations as useful for good governance (Bulkeley & Betsill, 2005; Rhodes, 2007). The constructivist epistemological lens further reveals how seemingly apolitical issues such as the evidence used in decision-making and governance institutions and actors are in fact highly politicised, and that institutional norms, cultures and practices shape our ability to define and resolve societal issues and devise governance solutions.

The theory of reflexive governance stipulates inclusivity, the ability to embrace complexity and the scope for learning in governance practices (Voß & Kemp, 2006). Reflexive governance aims to achieve societal ends while avoiding unintended effects through broad, participatory problem-solving approaches. It does not seek predetermined solutions but supports the exploration of a broad set of decision alternatives with respect to a diverse set of criteria. Learning is an ongoing quest, and therefore, reflexive governance provides platforms for interaction that complement conventional political decision-making. It evaluates and reconsiders societal ends against the background of diverse concepts and values and embraces experiments, such as participatory scenario planning, which support the reassessment of needs and interests or the identification of other ways of meeting them. It aims to actively involve diverse viewpoints in articulating problem definitions – even from actors who have limited capacities to represent themselves (avoiding non-decision-making).

References

Allen, J. (2011). *Topological twists*. SAGE Publications. https://doi.org/10.1177/2043820611421546

Arendt, H. (1986). Communicative power. In S. Lukes (Ed.), *Power* (pp. 59–74). New York University Press.

Bastos Lima, M. G., Persson, U. M., & Meyfroidt, P. (2019). Leakage and boosting effects in environmental governance: A framework for analysis. *Environmental Research Letters*, *14*(10), 105006. https://doi.org/10.1088/1748-9326/ab4551
Beck, U. (2006). Reflexive governance: Politics in the global risk society. In J. P. Voss, D. Bauknecht, & R. Kemp (Eds.), *Reflexive governance for sustainable development* (pp. 31–56). Edward Elgar Publishing. https://doi.org/10.4337/9781847200266.00011
Bellosta-Diest, A., Campo-Bescós, M. Á, Zapatería-Miranda, J., Casalí, J., & Arregui, L. M. (2022). Evaluation of nitrate soil probes for a more sustainable agriculture. *Sensors*, *22*(23), 9288. https://doi.org/10.3390/s22239288
Bevir, M. (2011). Governance as theory, practice, and dilemma. In M. Bevir (Ed,), *The Sage handbook of governance*. SAGE Publications Ltd. https://doi.org/10.4135/9781446200964.n1
Biermann, F., & Kim, R. (2020). Architectures of earth system governance: Setting the stage. In F. Biermann & R. Kim (Eds.), *Architectures of earth system governance: Institutional complexity and structural transformation* (pp. 1–34). Cambridge University Press.
Bohman, J. (1999). International regimes and democratic governance: Political equality and influence in global institutions. *International Affairs*, *75*(3), 499–513. https://doi.org/10.1111/1468-2346.00090
Boswell, J., Dean, R., & Smith, G. (2023). Integrating citizen deliberation into climate governance: Lessons on robust design from six climate assemblies. *Public Administration*, *101*(1), 182. https://doi.org/10.1111/padm.12883
Brousseau, E., Dedeurwaerdere, T., & Siebenhuner, B. (2012). Introduction. In E. Brousseau, T. Dedeurwaerdere, & B. Siebenhuner (Eds.), *Reflexive governance for global public goods*. The MIT Press.
Bulkeley, H. (2012). Governance and the geography of authority: Modalities of authorisation and the transnational governing of climate change. *Environment and Planning A*, *44*(10), 2428–2444. https://doi.org/10.1068/a44678
Bulkeley, H., & Mol, A. P. J. (2003). Participation and environmental governance: Consensus, ambivalence and debate. *Environmental Values*, *12*(2), 143–154. https://doi.org/10.3197/096327103129341261
Bulkeley, H., & Betsill, M. (2005). Rethinking sustainable cities: Multilevel governance and the 'Urban' politics of climate change. *Environmental Politics*, *14*(1), 42–63. https://doi.org/10.1080/0964401042000310178
Cairney, P. (2020). *Understanding public policy: Theories and issues* (2nd ed.). Palgrave Macmillan.
Castells, M. (2004). *The network society: A cross-cultural perspective*. Edward Elgar Publishing Limited.
Dedeurwaerdere, T., Brousseau, E., & Siebenhuner, B. (2012). Conclusions. In E. Brousseau, T. Dedeurwaerdere, & B. Siebenhuner (Eds.), *Reflexive governance for global public goods* (pp. 315–320). The MIT Press.
Driscoll, D. A. (2023). Populism and carbon tax justice: The yellow vest movement in France. *Social Problems, 70*(1), 143–163. https://doi.org/10.1093/socpro/spab036
Eising, R. (2015). Multilevel governance in Europe. In J. Magone (Ed.), *Routledge handbook of European politics* (pp. 165–183). Routledge Handbooks Online. https://doi.org/10.4324/9781315755830.ch9
Evans, J. P., & Thomas, C. (2024). *Environmental governance* (2nd ed.). Routledge.
Farrell, H., & Héritier, A. (2004). Interorganizational negotiation and intraorganizational power in shared decision making: Early agreements under codecision and their impact on the European Parliament and Council. *Comparative Political Studies*, *37*(10), 1184–1212. https://doi.org/10.1177/0010414004269833

Folke, C., Polasky, S., Rockström, J., Galaz, V., Westley, F., Lamont, M., Scheffer, M., Österblom, H., Carpenter, S. R., Chapin, F. S., Seto, K. C., Weber, E. U., Crona, B. I., Daily, G. C., Dasgupta, P., Gaffney, O., Gordon, L. J., Hoff, H., Levin, S. A., ... Walker, B. H. (2021). Our future in the Anthropocene biosphere. *Ambio*, *50*(4), 834–869. https://doi.org/10.1007/s13280-021-01544-8

Grin, J. (2006). Reflexive modernisation as a governance issue or designing and shaping re-structuration. In VossJ. , D. Bauknecht, R. Kemp, J. Vob, & C. R. Mccann (Eds.), *Reflexive governance for sustainable development* (pp. 57–81). Edward Elgar Publishing Limited.

Haas, P. (2004). When does power listen to truth? A constructivist approach to the policy process. *Journal of European Public Policy*, *11*(4), 569–592. https://doi.org/10.1080/1350176042000248034

Ham, C., & Hill, M. (1993). *The policy process in the modern capitalist state.* Harvester Wheatsheaf.

Hooghe, L., & Marks, G. (2004). Contrasting views of multi-level governance. In I. Bache & M. Flinders (Eds.), *multi-level governance* (pp. 15–30). Oxford University Press.

ICLEI. (2025). *Our approach: Sustainable urban development is key to a more sustainable world.* ICLEI Local Governments for Sustainability. Retrieved September 24, 2025, from https://iclei.org/our_approach/

Izcara Palacios, S. (1998). Farmers and the implementation of the EU nitrates directive in Spain. *Sociologia Ruralis*, *38*(2), 144–162. https://doi.org/10.1111/1467-9523.00069

Jackson, T. (2025). *The care economy*. Polity.

Juntti, M. (2012). Implementing cross compliance for agriculture in the EU: Relational agency, power and action in different socio-material contexts. *Sociologia Ruralis*, *52*(3). https://doi.org/10.1111/j.1467-9523.2012.00564.x

Juntti, M., & Potter, C. (2002). Interpreting and reinterpreting agri-environmental policy: Communication, trust and knowledge in the implementation process. *Sociologia Ruralis*, *42*(3), 215–232.

Kaljonen, M. (2006). Co-construction of agency and environmental management. The case of agri-environmental policy implementation at Finnish farms. *Journal of Rural Studies*, 22(2), 205–216. https://doi.org/10.1016/j.jrurstud.2005.08.010

Kuokkanen, A., & Yazar, M. (2018). Cities in sustainability transitions: Comparing Helsinki and Istanbul. *Sustainability (Basel, Switzerland)*, *10*(5), 1421. https://doi.org/10.3390/su10051421

Latour, B. (2004). *Politics of nature: How to bring the sciences into democracy.* Harvard University Press.

Lukes, S. (2017). *Power: A radical view*. Blackwell.

Lundy, D. (2017). *Lobby planet Brussels: The Corporate Europe Observatory guide to the murky world of EU lobbying.* Corporate Europe Observatory. https://corporateeurope.org/sites/default/files/lp_brussels_report_v7-spreads-lo.pdf

Mcdermott, T. K. J., & Surminski, S. (2018). *How normative interpretations of climate risk assessment affect local decision-making: An exploratory study at the city scale in Cork, Ireland.* The Royal Society. https://doi.org/10.1098/rsta.2017.0300

Morrison, T. H., Adger, W. N., Brown, K., Lemos, M. C., Huitema, D., Phelps, J., Evans, L., Cohen, P., Song, A. M., Turner, R., Quinn, T., & Hughes, T. P. (2019). The black box of power in polycentric environmental governance. *Global Environmental Change*, *57*, 101934. https://doi.org/10.1016/j.gloenvcha.2019.101934

O'Neill, K. (2022). International organizations: Global and regional environmental cooperation. In P. Harris (Ed.), *Routledge handbook of global environmental politics* (pp. 105–117). Routledge.

Ostrom, E., Burger, J., Field, C. B., Norgaard, R. B., & Policansky, D. (1999). Revisiting the commons: Local lessons, global challenges. *Science*, *284*(5412), 278–282. https://doi.org/10.1126/science.284.5412.278
Oxfam. (2023). *Climate equality: A planet for the 99%*. Oxfam GB. https://doi.org/10.21201/2023.000001
Park, S. (2018). Constructivism. In T. G. Weiss & R. Wilkinson (Eds.), *International organization and global governance* (pp. 146–158). Taylor & Francis Group.
Parsons, T. (1986). Power and the social system. In S. Lukes (Ed.), *Power* (pp. 94–143). New York University Press.
Raworth, K. (2017). *Doughnut economics: Seven ways to think like a 21st century economist* (1st ed.). Random House Business Books.
Rhodes, R. A. W. (2007). Understanding governance: Ten years on. *Organization Studies*, *28*(8), 1243–1264. https://doi.org/10.1177/0170840607076586
Sabatier, P. (1986). Top-down and bottom-up approaches to implementation research: A critical analysis and suggested synthesis. *Journal of Public Policy*, *6*(1), 21–48. https://doi.org/10.1017/S0143814X00003846
Scoones, I., Stirling, A., Abrol, D., Atela, J., Charli-Joseph, L., Eakin, H., Ely, A., Olsson, P., Pereira, L., Priya, R., van Zwanenberg, P., & Yang, L. (2020). Transformations to sustainability: Combining structural, systemic and enabling approaches. *Current Opinion in Environmental Sustainability*, *42*, 65–75. https://doi.org/10.1016/j.cosust.2019.12.004
Sell, S. K. (2018). Who governs the globe. In T. Weiss & R. Wilkinson (Eds.), *International organization and global governance* (pp. 77–90). Taylor & Francis Group.
Sze, J. (2020). *Environmental justice in a moment of danger*. University of California Press.
Sørensen, E. (2002). Democratic theory and network governance. *Administrative Theory & Praxis*, *24*(4), 693–720. https://doi.org/10.1080/10841806.2002.11029383
Taylor, M. (2007). Community participation in the real world: Opportunities and pitfalls in new governance spaces. *Urban Studies*, *44*(2), 297–317. https://doi.org/10.1080/00420980601074987
Turnpenny, J., Lorenzoni, I., & Jones, M. (2009). Noisy and definitely not normal: Responding to wicked issues in the environment, energy and health. *Environmental Science & Policy*, *12*(3), 347–358. https://doi.org/10.1016/j.envsci.2009.01.004
UNCCD. (2025). *Land degradation neutrality fund*. United Nations Convention to Combat Desertification. Retrieved September 25, 2025, from https://www.unccd.int/land-and-life/land-degradation-neutrality/impact-investment-fund-land-degradation-neutrality
UNFCCC. (2022). *Climate finance in the negotiations*. Retrieved March 18, 2022, from https://unfccc.int/topics/climate-finance/the-big-picture/climate-finance-in-the-negotiations
UN General Assembly. (2015). *Transforming our world: The 2030 Agenda for Sustainable Development* (A/RES/70/1). United Nations.
Voß, J. P., & Kemp, R. (2006). Sustainability and reflexive government: Introduction. In *Reflexive governance for sustainable development*. Edward Elgar Publishing. https://doi.org/10.4337/9781847200266.00009
Watson, M. (2014). International political economy. In Telò (Ed.)M. , *Globalisation, multilateralism, Europe: Towards a better global governance?* (pp. 145–160). Taylor & Francis Group.
Weiss, T. G., & Wilkinson, R. (2018). *From international organization to global governance* (T. Weiss & R. Wilkinson, Eds., 2nd ed.). Routledge, Taylor & Francis Group.

Part 2

Understanding global governance issues and solutions

5 Governing the climate crisis

Learning objectives

- Which sectors of the economy are the most significant in terms of greenhouse gas (GHG) emissions and climate change drivers?
- What exactly is climate justice and how should it inform climate governance?
- What are the key features of climate mitigation policy, such as the Paris Agreement, and how are GHG emissions measured and reported?
- What does adaptation policy consist of and how does it function?
- What is the climate policy implementation gap and what are its main causes?

5.1 Introduction

Climate change is a defining issue of our time. It exacerbates other key environmental challenges, such as loss of biodiversity and land system change, and accelerates inequality as its impacts are particularly devastating for poor countries and communities (Folke et al., 2021; Steffen et al., 2015). Numerous scientific reports evidence rising greenhouse gas (GHG) concentrations and an associated rise in global temperature since the beginning of the industrial era (IPCC, 2022b, 2023; Stoller-Conrad, 2023). This warming is causing unpredictable and adverse weather patterns such as intensifying storms, drought, and flooding, risking livelihoods, food security, infrastructure, and buildings. Rising sea levels threaten the territory of numerous countries, and changes in ocean currents risk a significant departure from the climatic conditions stabilised during the Holocene (Folke et al., 2021; IPCC, 2022b; Yule et al., 2023). Climate change also has a direct effect on people's physical wellbeing. High temperatures and resulting increased levels of pollution compromise the health of vulnerable populations such as the very old and very young, and those living in sub-standard dwellings.

The United Nations Framework Convention on Climate Change (UNFCCC) initiated a global process of governing climate change in 1992. But it took until the Paris Agreement in 2015 for a first substantial commitment to GHG reduction when the global 'net zero' by 2050 target was

DOI: 10.4324/9781003179207-7

agreed (UNFCCC, 2024). Net zero, sometimes also referred to as zero carbon or carbon neutrality, refers to limiting GHG or CO_2 concentrations in the atmosphere by not emitting more than what can be absorbed by nature. But progress in mitigating emissions has been slow, and it seems that the Paris target of limiting warming to 1.5°C has already been breached. Indeed, following the Global Stocktake in 2023, it seems that the planet is on course for 1.8–2.5°C of warming, having crossed the 1.5° mark in 2024 (Darby et al., 2024; King & Cassidy, 2025).

Climate justice refers to the uneven manner in which the causes and impacts of climate change are distributed across the globe. Climate change and its impacts provide a case in point for Ulrich Beck's theory of Risk Society (Beck, 1992). The boomerang effect, collective risk and yet, unequal risk positions can be used to describe the way the impacts of climate change unfold unequally across geographies and socio-economic strata. While impacts of centuries of GHG emissions mainly by industrialised countries in the Northern hemisphere are most acutely felt near the Equator and large river deltas of developing countries, the indirect impacts on global supply chains are driving a mounting collective risk of destabilisation across the globe.

Most GHG emissions are caused by the energy sector (IPCC, 2023). Therefore, a transition to renewable, GHG neutral energy is urgently needed. While a 2023 global Conference of Parties (COP) produced the pledge to 'transition away from fossil fuels', progress in this has so far been too slow to curb the overall growth in emissions. The energy sector underpins most economic activity and therefore the political will to commit to such a transformational change has proven variable. But energy consumers from industry to transport and households should not sit back and wait for top-down intervention as there is an urgent need for technological development and behavioural change that will support the transition to renewables, cut energy use and change consumption habits.

Therefore, climate change and its impacts are not just a matter for scientific analysis. Social and cultural constructions of climate risks and conflicts of interest and trade-offs implicated by mitigation activities need to inform governance interventions (Beck, 1996). Climate change impacts touch individuals and communities in many different ways, from displacement caused by flooding or drought to climate anxiety impacting particularly young people (Whitmarsh et al., 2022). As most socio-environmental issues, climate change is characterised by a degree of uncertainty, differing risk perceptions, conflicting interests, and fluctuating political will, which renders it a highly politicised, wicked governance issue (Turnpenny et al., 2009). Governance needs to urgently curb the steady progress of global warming while dealing with highly unequal climate-related drivers and risk burdens that often align with socio-economic disadvantage both globally and locally. In this context, formulating legitimate interventions that mitigate GHG emissions and support adaptation to the already inevitable impacts requires careful engagement with the different aspects of justice discussed in Chapter 3.

In line with other chapters of the book, this chapter considers governance to mean *the institutions and practices of decision-making and intervention regarding issues of public interest that are undertaken by formal and informal governance actors, often situated in networked relations of power* (Morrison et al., 2019; Sørensen, 2002). This encompasses a broad range of interventions to address climate change, and examples have been chosen to illustrate the plethora of approaches at global and local scales, driven by either formal or informal governance actors.

5.2 The drivers of the climate crisis

In quantitative terms, Carbon Dioxide (CO_2) is the main greenhouse gas (GHG) contributing to the warming of the Earth's atmosphere. Other gases with a similar impact are water vapour, Nitrous Oxides (NOX), Methane (MH4), and Chlorofluorocarbons (CFCs). The human-induced increase in the concentration of these gases in the atmosphere causes the 'greenhouse effect' by letting radiation from the sun enter but keeping the heat from escaping (IPCC, 2007).

Scientific evidence plays a crucial role in our understanding of, ability to mitigate and preparedness for climate change. The International Panel on Climate Change (IPCC) is the main scientific advisory body informing the UN Framework Convention to Combat Climate Change and its signatory states. The IPCC consists of a large group of scientists from all over the world, based in a range of universities and research institutions, collaborating in three Working Groups:

Working Group I – The Physical Science Basis,
Working Group II – Impacts, Adaptation and Vulnerability, and
Working Group III – Mitigation of Climate Change.

The IPCC produces regular Assessment Reports, which collate scientific findings regarding the progress of climate change and its drivers and impacts. These Assessment Reports have played a fundamental role in informing and furthering global climate policy in the past two decades. Usually, climate change causing emissions are expressed in 'carbon equivalent' or CO_{2e} regardless of what gas they consist of. For the purposes of tracking progress in emission reduction commitments made since the Paris Agreement, the focus is on annual emissions of CO_{2e} by country (Ritchie et al., 2020). To account for the varying population sizes, this is often broken down to emissions per person (CO_{2e} per capita).

CO_{2e} emissions have continued to rise throughout the industrial era, and, despite committments such as the Paris Agreement in 2015, the decade from 2010 constitutes the decade of highest ever emissions (Figure 5.1). The largest growth in absolute emissions occurred in CO_2 from fossil fuels and industry followed by CH_4. In 2019, approximately 34% of total net anthropogenic CO_{2e} emissions came from the energy supply

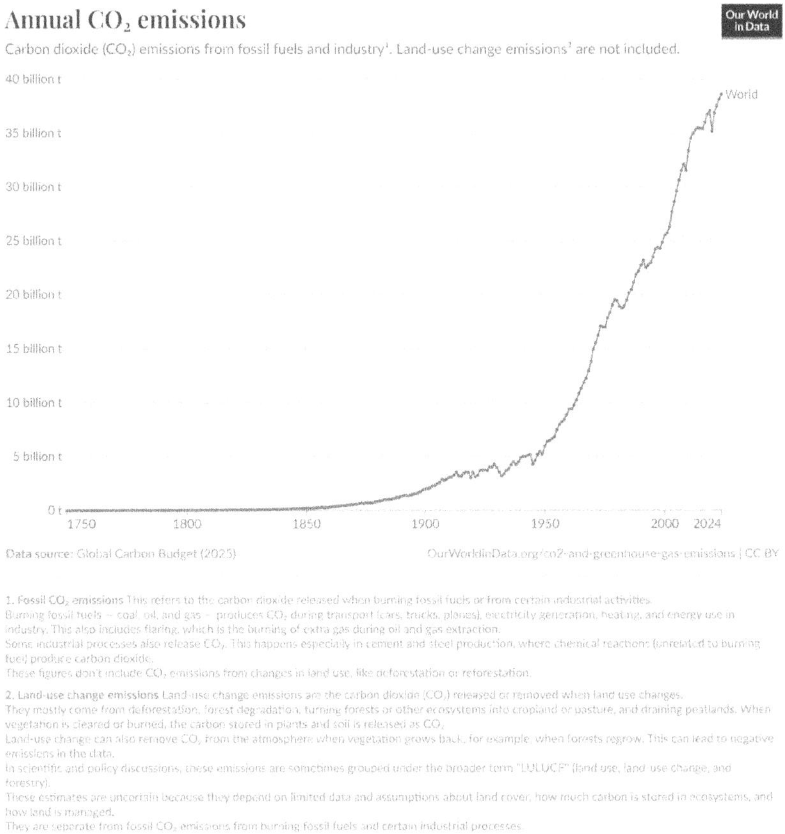

Figure 5.1 Annual CO_2 emissions from fossil fuels and industry since 1750 (Ritchie, H., 2026).

Data Source: Global Carbon Budget (2025) – with major processing by Our World in Data.

sector, 24% from industry, 22% from agriculture, forestry and other land uses, 15% from transport and 6% from buildings (IPCC, 2022a). While the largest individual subsector contributing to global CO_{2e} emissions was electricity and heat generation (implicating the energy sector but also industry and buildings), the fastest growth of emissions can be observed in transport, particularly aviation. The International Energy Agency (IEA) reports that in 2024, energy demand grew by 2.2% and energy sector (actual) CO_2 emissions reached a new record (IEA, 2025).

This gives us some idea of where we need to concentrate mitigation efforts in terms of reducing activity, making efficiency savings and transferring to non-emitting activities such as energy production from renewable, carbon-neutral sources. Achieving net zero requires retrofitting old housing stock and making new buildings carbon neutral, ending mass ownership of combustion-engine cars, expanding zero-carbon public transport, electrifying industries, rolling out a zero-carbon energy supply and, crucially, phasing out fossil fuel production (Anderson, 2023). While overall emissions are steadily rising, the *growth* in annual emissions is beginning to slow (IEA, 2025; IPCC, 2022a). This speaks to the ability of our societies and economies to adjust in response to the scientific evidence of potentially disastrous warming, but crucially, so far at an insufficient pace (see Chapter 3 for the discussion on decoupling – the 2025 report from IEA shows that progress on decoupling has slowed down in the past couple of years).

The largest economies of the world have begun to invest in clean technologies and have managed to mitigate their emissions. There has been good progress in shifting to renewable energy sources in rich countries in particular. Although energy demand continued to grow, as shown in Figure 5.2, in 2024, renewables accounted for the largest share of the growth in total energy supply (38%), followed by natural gas (28%), coal (15%), oil (11%) and nuclear (8%) (IEA, 2025). Nevertheless, in terms of historic emissions, rich countries stand out as top emitters, therefore holding the greatest responsibility for causing climate change (Darby et al., 2024). According to Our World in Data, the US is the largest contributor to cumulative emissions since 1750, accounting for nearly a quarter. The EU of 27 states is the second largest, with responsibility for a fifth of cumulative global emissions (Ritchie & Roser, 2024).

Crucially, Richie et al. (2020) point out that a country's CO_{2e} is formally calculated from territorial emissions, which excludes the emissions related to the consumption of goods that have been imported from other

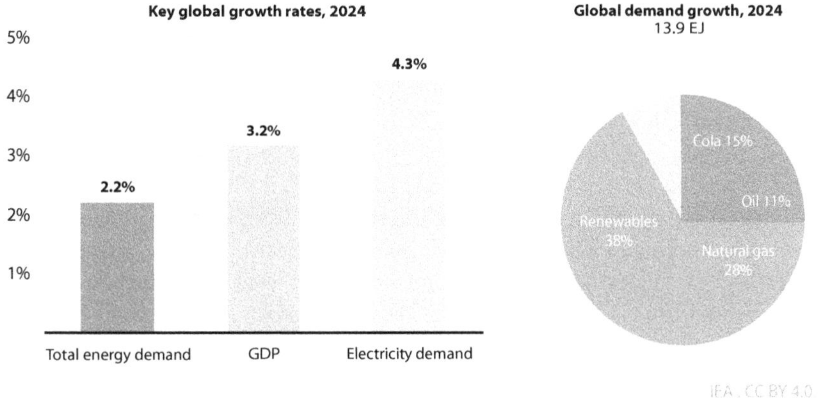

Figure 5.2 Growth rates in total energy demand, GDP and electricity demand, and the growth share of different energy sources in 2024. (adapted from IEA2025 under the IEA CC BY 4.0 licence).

This is a work derived by Meri Juntti from IEA material, and Meri Juntti is solely liable and responsible for this derived work. The derived work is not endorsed by the IEA or its Member countries in any manner.(adapted from IEA2025 under the IEA CC BY 4.0 licence).

countries. Therefore, many of the emissions from our material consumption are not taken into consideration in the national emission inventories, which all the signatories to the UN FCCC and the Paris Agreement have to produce on 2-yearly basis (Barrett et al., 2013a). There is no doubt that the 'convenience lifestyles' relying on carbon-intensive consumer goods prevalent in the so-called developed countries continue to carry significant responsibility for global GHG emissions.

To sum up, we know that human activities have significantly changed the constitution of the Earth's atmosphere and are the driving force of the warming that we are witnessing presently (IPCC, 2007). However, the different countries of the world are in a very unequal position in relation to the climate crisis. The so-called developed countries like the US, UK, and EU Member States hold much more responsibility for emissions than others. Despite recent advances in curbing emissions of production-based activities in these countries, much more substantial efforts to decarbonise industry, dwellings and transport across the world are needed, and this requires the transfer of both finance and technology from the rich to poor countries.

5.3 Inequality and climate change

5.3.1 Climate justice

Climate justice draws attention to the unequal responsibility for GHG emissions, unequal risk burden resulting from climate change impacts and unequal ability to implement both mitigation and adaptation measures across the world (Barrett, 2012; Coolsaet, 2020). Due to their geographical location and level of infrastructural and economic development, the so-called developed countries tend to be much less vulnerable to the unfolding impacts of the climate crisis than poorer, developing countries.

Regions at heightened risk are the Arctic, because of the impacts of high rates of projected warming on natural systems and human livelihoods; Africa, because of low adaptive capacity and the scale of projected climate change impacts; small island states, where there is high exposure of population and infrastructure to drastic impacts such as sea-level rise; and Asian and African mega-deltas, due to large populations and high exposure to sea-level rise, storm surges and river flooding (IPCC, 2014). So, while developed countries are the biggest emitters and the most resilient, developing countries hold the least responsibility for climate change but are the most affected by its impacts. This situation constitutes a global scale climate injustice (Barrett, 2012; Coolsaet, 2020).

For example, a large industrialised country like the USA will have caused manifold more emissions than the small island state of Fiji (Ritchie & Roser, 2024). Focusing on per capita emissions, an average US citizen emits 14.93 tonnes per annum and a Fijian only 1.17 tonnes (Ritchie et al., 2020). But Fiji is an example of a small island state that is already acutely impacted by climate change (Oxfam, 2023). Sea-level rise threatens the very existence of Fijian territory and has already displaced communities and compromised peoples' ability to make a living (IPCC, 2022b). Many suggest that it is therefore very unfair and indeed unacceptable to require Fiji to make comparable sacrifices for emission mitigation as the US.

Restorative justice (McCauley & Heffron, 2018) makes a moral argument for action by those responsible for the majority of emissions to compensate those who are most vulnerable and exposed to their impacts. It has been evoked in global negotiations to oblige leading emitters to fund both mitigation and adaptation measures in countries that remain vulnerable to climate crisis due to inequalities that date back to the colonial era. So-called restitution payments or compensation for loss and damage have been subject to intense negotiations and at the COP 28 in 2023 in Dubai, an agreement was reached on the Loss and Damage Fund. This will hopefully see the transfer of funds from rich countries to the Least Developed Countries, small island states and others who are suffering from the worst impacts of climate change and need help with adaptation and repair (UNFCCC COP28, 2023).

But by and large, climate justice remains a contentious and overlooked issue in global climate negotiations, and pledged funds have been slow to deliver. In an effort to encourage private investment to address this gap, Just Energy Transition Partnerships (JETPs) have been established among select developing countries and their more developed partners (UNRISD, 2023). A key aim of this governance instrument is to speed up the energy transition to renewables to aid the net zero transition in developing countries. JETPs constitute investment plans that will direct money from public and private entities in the developed countries to their poorer partners.

Just transition (Oxfam, 2023) addresses climate justice from the perspective of the so-called 'mitigation burden'. It pays attention to the impact of emission mitigation policies on poor countries and households, and employment sectors that tend to be adversely affected not only by climate change impacts but also by mitigation policies such as carbon taxes, subsidy removals or high energy prices. McCauley and Heffron (2018, p. 2) define just transition as 'a fair and equitable process of moving towards a post-carbon society. This process must seek fairness and equity with regards to the major global justice concerns such as (but not limited to) ethnicity, income, gender within both developed and developing contexts'. For example, the fossil fuel energy sector is a huge employer globally and many more industries and jobs depend on access to cheap energy. A just transition will ensure that these people are guaranteed an income and an opportunity for re-training as the fossil fuels are phased down. Restorative justice and governance instruments (such as subsidies and the provision of re-education for new careers) are central to just transition.

5.3.2 Systemic drivers of climate injustice

Ulrich Beck (1996) terms climate change a 'deep institutional crisis of ... industrial modernity' (Beck, 1996, p. 12). He refers to the way that fossil-based energy sources, technological development and Western science were harnessed to deliver development as modernisation by the powerful colonial countries over the era of industrialisation (often deemed to have started mid-18th century). This agenda became institutionalised in the still prevalent 'modern' structures of knowledge production, societal organisation and the global economy. Beck argues that it is these institutions that now underpin the current inequalities and the planetary crisis.

For example, the discovery of fossil energy was a crucial factor in the initiation of the growth-oriented capitalist market economy (Clark & Szerszynski, 2020). Institutional investors, banking systems, and energy companies were established early in the industrial era to enable and regulate access to the underground energy resources. The impact of fossil fuels on the power structures of today's society is therefore immense (Jackson, 2017). Not only does the capitalist market economy continue to accumulate wealth in the hands of capital owners (Piketty, 2014) but for example, urban geographers describe how our cities can be seen as manifestations of surplus industrial revenue in need of investment and are therefore shaped by the same rationale of revenue accumulation (Harvey, 2006). Similarly, our modern convenience lifestyles are informed by constantly emerging new 'needs' that can only be satisfied by consumption, generating revenue for the economy. We are all embroiled in fossil fuel-driven economies whose institutions are designed to safeguard access to capital and to facilitate the extraction of profit, and on which we now depend for survival.

Similarly, colonial power relations became institutionalised thanks to these developments. Present-day economic and climate injustices therefore mirror colonial relations (Ghosh, 2021). Efforts for climate justice are not limted to addressing the inequality in responsibility for emissions and exposure to attendant risks but also take issue with the fundamentally unjust societal structures that are upheld by and poised to serve the so-called fossil capitalism (Clark & Szerszynski, 2020; Huber, 2009). Some examples of financial institutions that favour rich developed countries and perpetuate global inequality, such as credit rating agencies and varied interest rates for national debt, were discussed in Chapter 2.

Beck (1996) suggests that, in this context, efforts to govern climate change through incremental mitigation and adaptation measures exemplified by the green economy agenda (Chapter 3) instils a false sense of control. Beck terms it 'placation' rather than effective action. Instead of addressing the drivers of cliamte change and climate injustice, incremental governance interventions serve to legitimise societal processes – energy solutions and industrial activities as well as our prevalent consumption-oriented lifestyles – which drive climate change (IPCC, 2007, 2023). Instead of addressing the fundamental drivers of climate-related risks vested in the growth-oriented fossil economy, current mitigation efforts are focused on living with climate risks, which are portrayed as inevitable and manageable. Anderson (2023) argues that this conciliatory framing of climate change is evident even in the IPCC's reports, which underpin for example the 2050 deadline for global net zero. He argues that to avoid catastrophic warming, most ‘developed’ nations need to reach zero CO_2 emissions between 2030 and 2035, with developing nations following suit up to a decade later (Anderson, 2023). This requires a much more rapid shift away from fossil-based energy than what has so far been instigated.

In addition to the growth-based economic system, another key obstacle to progress in GHG mitigation is the misalignment between experienced climate risk and the need for behavioural change. In the parts of the globe where GDP per capita is high and spending power among populations is strong, climate change impacts remain less prevalent and less threatening to everyday life activities and quality of life. The largest emitters, the so-called developed economies, are therefore fertile ground for problematic cognitive dissonance where awareness of the climate crisis does not lead to concrete actions to address it (Thierry et al., 2023). The crisis is felt more acutely in countries where the per capita emissions are much smaller than in the US or indeed in most European countries. Poorer populations experience much more severe risks and have less access to risk management devices such as flood defences and insurance. Indeed, the fear of what may happen to Western convenience-oriented high-consumption lifestyles and aspirations in a carbon-neutral, post fossil-fuel world may constitute a key barrier to urgently needed progress.

The degrowth pathways to sustainability discussed in Chapter 3 of this volume involve efforts to instigate a cultural shift away from consumption-oriented values and norms. Degrowth solutions are promoted by many as safer and fairer routes to net zero than ones relying on novel, expensive and complex technological solutions such as carbon capture that only rich countreis are likely to have access to (Keyßer & Lenzen, 2021). But it is difficult to summon political will for degrowth, a concept which so explicitly contradicts the dogma of the capitalist market economy and its powerful beneficiaries. There are some signs that this resistance may be softening. For example, in 2018 and 2023 the European Parliament organised a Post-Growth Conference to share progress in understanding alternatives to the growth-oriented capitalist economy as a blueprint for societal organisation towards sustainability. But the economic elite who benefit from the status quo – even as the vast majority of the global population are experiencing a decrease in living standards and the quality of life due to climate change and growing inequality – are still reluctant to look to alternatives.

The next section explores existing and planned governance solutions addressing climate mitigation and adaptation from a critical perspective with a view to their ability to support a just and effective transition to a carbon-neutral society to avert the worst consequences of the mounting climate breakdown.

5.4 Climate change governance

Climate change constitutes a global issue that no state can meaningfully address alone. Therefore, the UN Framework Convention on Combatting Climate Change (UNFCCC Box 4.1 in Chapter 4) was established in the wake of the Earth Summit in 1992. It forms the backbone for global governance for climate change. Nation states are motivated to form multilateral agreements due to the benefits vested in interstate cooperation, better guarantees that other states are also complying with commonly agreed targets and rules, and because membership of multilateral agreements provides an opportunity to influence forthcoming regulation (Sell, 2018). Broadly speaking, global climate governance has two aspirations: emissions reduction, aimed at minimising the causes of climate change (mitigation policy), and increasing resilience towards the already inevitable impacts of climate change (adaptation policy). This section will discuss the instruments and challenges of climate governance from the perspective of both mitigation and adaptation.

5.4.1 Mitigation policy

The Paris Agreement in 2015 (UNFCCC, 2024), set out the global goal of limiting global warming to 2°C or, where possible, to 1.5°. To this end, national governments, cities, organisations, and businesses must measure, reduce, and where necessary, offset greenhouse gas (GHG) emissions emanating from their territories and/or activities (UNFCCC, 2024). This must be done in line with the global GHG Protocol for measuring and accounting GHG emissions expressed as carbon dioxide equivalent (CO_{2e}). The Protocol is globally certified by the ISO 14064 standard and enables accurate assessment of global GHG emissions by ensuring that countries provide standardised and comparable reports. As a part of the Protocol, bi-annual Net Inventory Reports (NIRs) are to focus on three types of emissions:

—Scope 1: direct emissions of CO_{2e} resulting from operations of the organisation.
—Scope 2: indirect emissions of CO_{2e} resulting from the production of electricity and heat for the operations of the organisation.
—Scope 3: indirect emissions of CO_{2e} produced by others but associated with the organisation in question.

NIRs must encompass all emissions from state territories (territorial emissions), while emissions from aviation and shipping in international territories are noted as memos. The Protocol does not extend to the emissions caused by the consumption of goods, which often add significantly to the emissions produced by the populations of so-called developed or rich countries (Barrett et al., 2013b; Ritchie et al., 2020).

Since 2020, the signatories to the Paris Agreement have devised national climate action plans (or Nationally Determined Contributions, NDCs) detailing five-year mitigation targets, how they intend to achieve them, and what kind of adaptation measures will be undertaken to increase resilience (UNFCCC, 2024). The NDCs are monitored via the NIRs and may be informed by Long Term Emission Development Strategies (LT-EDS), which provide the Parties to the Paris Agreement an opportunity to set out how their emission targets align with a broader range of national development targets and plans. The 'ratchet mechanism' demands that each five-year NDC improves on the previous, including more ambitious mitigation targets and measures ensuring timely progress towards net zero. While the NDCs are mandatory, the Parties can draw up LT-EDS at their discretion.

The Paris Agreement also sets out a framework for capacity building support for climate mitigation and adaptation in developing countries (UNFCCC, 2024). There is therefore a recognition of restorative justice, whereby developed countries need to take the lead and provide support for those that are lagging behind. The support should take the form of interventions such as financial assistance from the Green Climate Fund; technology transfer increasing access to cleaner technological solutions such as renewable energy or more carbon-efficient production systems; or capacity building activities such as training in low-carbon solutions and institutional strength that are crucial for good governance.

NDCs rely on a mix of economic, regulatory and informational governance instruments to realise the cuts in emissions. These may encompass bans or taxes on emission-intensive products such as disposable plastic goods, subsidies to support energy efficiency improvements and the transition to renewables, or cap-and-trade systems whereby only a sanctioned amount of GHG emissions is allowed free of cost (Evans & Thomas, 2024). The Kyoto Protocol, passed by the UNFCCC in 1997 and operationalised from 2005 to 2020, was the first example of a multilateral cap-and-trade GHG mitigation scheme. The Kyoto Protocol was adopted by 37 developed countries, which agreed to adopt emission quotas – a sanctioned amount of GHGs – that were determined on the basis of country-specific 'needs' (which constituted a much-criticised feature of the Protocol). The quotas consisted of a number of carbon credits that the countries could then trade among each other. While the quota set a cap on overall emissions, if a country was able to limit its emissions below its quota, it could make an income from selling the remaining credits to a country that was unable to meet its respective quota. In this way, a market for carbon emissions was created, whereby the price of carbon credits would serve as a 'penalty' for those who exceeded their quotas (Evans & Thomas, 2024). It also meant that a country could recover some of the cost of investing in clean technology from the sale of its remaining carbon credits. The Kyoto Protocol also included the Clean Development Mechanism (CDM), enabling the 37 signatories to offset their own GHG emissions by investing in clean technologies in developing countries.

The Paris Agreement marked a shift from the top-down allocation of emission quotas to a bottom-up system of nationally determined emission targets and NDCs (Harris, 2022). The ambition for emission mitigation is now determined at the national scale, which some find problematic as it risks slowing down progress. But COP26 in Glasgow in 2021 achieved a significant step forward in the employment of market-based governance interventions for climate mitigation (Evans & Thomas, 2024). COP26 saw the Parties to the Paris Agreement agree a road map towards the realisation of a full transnational carbon market that has the potential to incentivise and fund an overall mitigation in global emissions (Minas, 2022). While the Kyoto Protocol pertained to 37 'developed' countries and integrated developing countries through the Clean Development Mechanism, a transnational market will hopefully soon enable all countries to trade carbon in a manner that incentivises and supports the fulfilment of all NDCs.

Nevertheless, so far mitigation governance has failed to achieve required progress on the Paris targets. The 28th COP held in Dubai in 2023 was pivotal in that it published a Global Stocktake of progress in emission mitigation resulting from the Paris Agreement. This revealed a substantial implementation gap between the NDCs and the targets set to limit warming to 2°C (Darby et al., 2024). For example, China, the United States, India, the EU and Saudi Arabia were all lagging in a switch from fossil fuels to renewables. Whereas only Costa Rica, Denmark, France, Sweden and some small island states had set out firm plans to end investment in fossil fuels. Spurred by this evident implementation gap, the Stocktake included the historic pledge to begin a 'transition away from fossil fuels'. Over a hundred states also signed a pledge to triple existing renewable energy capacity to at least 11,000 GW by 2030 and to double the global average annual rate of energy efficiency over the same time period (UNFCCC COP28, 2023).

There have been calls to facilitate the engagement of private investment capital in support of the transition to renewable energy. To this end, so-called mini-multilateral Just Energy Transition Partnerships (JETPs) were set up outside the Paris Agreement to improve the potential for just transition (UNRISD, 2023). However, JETPs have been criticised for lack of transparency. Since they operate outside the Paris Agreement reporting mechanisms, it is not clear how exactly the participating developing countries are selected and what conditions are in place to ensure that the funded energy initiatives really support just and transformative change that addresses energy access as well as climate impact (UNRISD, 2023). Moreover, the JETPs favour conditional loans over grants, and this may prioritise solutions that provide fast economic returns at the expense of serving poor communities and regions where energy access is lacking but where there is little scope for high returns. Therefore, rather than empowering, loans may serve to build dependency.

5.4.1.1 Governance theories for analysing mitigation policy

The workings of global scale governance initiatives can be understood through the lens of intergovernmental regimes theory (Weiss & Wilkinson, 2018), which draws attention to the construction of shared norms and differences in power and interests across the parties to multilateral agreements. The UNFCCC operates based on unanimous decision-making, which means that national interests can influence collaboration and slow down decision-making (Darby et al., 2024). With the shift from top-down governance underpinning the Kyoto Protocol where the national quotas were centrally set, the NDCs mark a shift towards a nationally led bottom-up model of governance which is even more vulnerable to national interests (Harris, 2022). Many authors have used the multilevel governance approach and constructivist methodologies to understand the implementation gap plaguing mitigation efforts (Hooghe & Marks, 2004). Global commitments to climate mitigation targets and plans are enshrined in national legislation that then directs and informs administrations at a regional and local scale (Bulkeley & Betsill, 2005).

Bailey (2017) provides an example from Australia, where the Clean Energy Act came into force in 2011. The Act established the Clean Energy Mechanism (CEM), which priced carbon emissions at AU$23 per tonne with the intention of transitioning to a flexible market-based system later. However, the CEM was short-lived. It faced strong political party and industry opposition and was ultimately repealed in July 2014 (Bailey, 2017). The planned carbon emissions trading scheme was successfully contested on the grounds that it endangered livelihoods at the local scale, with particularly strong opposition from the agricultural lobby. The Australian example is not alone and speaks to the challenge of legitimacy in governing a complex wicked issue such as climate change. It also reveals differences in scalar framings of climate justice that come to the fore in multilevel governance (Bailey, 2017). While global climate justice frames Australia as a high emitter, at the local scale climate mitigation is pitted against access to income, livelihoods, and economic viability of key sectors such as, in this case, agriulture.

A policy that fails to gain legitimacy and risks adverse impact on business growth may lead to spill-over effects, where high emission businesses will transfer to other countries or regions where regulation is less stringent (Biermann & Kim, 2020). International coordination and linkages between sectoral policies are crucial for addressing this. The proposed emissions trading scheme in Australia (Bailey, 2017) preceded the 'Glasgow consensus' on a global GHG market. This may have contributed to its political resistance as the loss of business to other coutnries was seen as a threat to the national economy.

Biermann and Kim (2020) emphasise the complex problem-solving capacity of governance interventions, their ability to respond to unexpected problems and to maintain legitimacy, accountability and equity at a local scale where a broad range of policies and stakeholders coalesce. Democratic and transparent procedures of decision-making and implementation are crucial to achieving this, as well as the ability to resolve any uneven regulatory burdens. It is important to maintain sights on the threat to the long-term economic outlook posed by climate change in order not to fall victim to economic short-termism. There is a need to build viable just transition pathways such as compensation and access to alternative livelihoods into mitigation policies (Oxfam, 2023). This makes for a strong argument for an active search for degrowth alternatives, for example in the form of public investment in climate-friendly agriculture and other similar causes that also create viable employment opportunities as suggested in Chapter 3.

As mitigation ambitions are increasingly determined at national level following the Paris Agreement (Harris, 2022), it is important to recognise the agency of non-state actors in climate governance. This can be done with the help of network governance theories (Sørensen, 2002), which capture governance efforts that fall outside the hierarchical multilevel governance framework and a top-down command-and-control approach. This perspective provides a more nuanced understanding of climate governance, for example, in the 'Trump-2' era. The US President's second term has seen devastating cutbacks on domestic climate legislation, finance and reporting in its first year (Bomberg, 2025). Internationally, the US has yet again pivoted away from agreements and funding commitments; though fortunately, this time other countries have not followed suit to the same extent as in 2016. A focus on state scale, where the benefits of many of the so-called green economy policies are felt more directly in terms of employment and investment opportunities (Chapter 3), reveals that several projects and legislative initiatives still persist. Beyond the US, the Cities for Climate Protection (CCP) Programme operating as a part of the ICLEI Local Governments for Sustainability network demonstrates that local authorities can show significant leadership in climate governance (Bulkeley & Betsill, 2005). While there is no time to slow down action on climate change mitigation, successful local or state-level initiatives may in time help re-kindle national policy commitments.

5.4.2 Adaptation policy

Given that the world has likely now surpassed 1.5°C of warming and is on course to 1.8–2.5°C, climate change adaptation is imperative. Adaptation refers to 'actions to adjust to the new climatic conditions by taking appropriate measures and taking advantage of climate change benefits, if any' (IPCC, 2022b). It involves practical steps, such as planning for sea-level rise, managing water security, and integrating climate risks into infrastructure and policy (Ray Biswas & Rahman, 2023). Climate change adaptation is therefore focused on assessing and building resilience, which is a term stemming from ecological systems science but now widely modified to support theories of societal ability to withstand the effects of climate change.

The Global Goal on Adaptation was established as a part of the Paris Agreement in 2015 and seeks to protect people, livelihoods and ecosystems, particularly in vulnerable communities (UNFCCC, 2024). The goals seek to ensure that National Adaptation Plans are in place in every country and call for the transfer of finance from developed to developing countries to support place-based adaptation measures (Darby et al., 2024). Further progress was made when 82 governments endorsed the Declaration on Climate Relief, Recovery and Peace at COP28, committing themselves to enhance finance and planning for adaptation and resilience, though affordable finance for LDCs is yet to be realised at scale.

Resilience can be taken to mean 'the amount of change a system can undergo and still retain the same function and structure while maintaining options to develop' (Nelson et al., 2007): 396). But Evans & Thomas (2024) suggest that central to the resilience of any societal entity such as a community or a city is its capacity to modify system structure according to need. This supports a more politicised interpretation of climate change resilience as either transition or transformation, both of which address the driving forces of climate vulnerability (Pelling, 2012). Adaptation as transition mitigates vulnerability, for example, by allocating public funds to flood damage compensation. Adaptation as transformation, on the other hand, refers to a more fundamental reshaping of societal structures, norms, and institutions to tackle climate change vulnerabilities at their root and provide long-term sustainability rather than just mitigation of immediate risks. Transformation would involve legislative changes to alleviate inequality by prioritising wellbeing and community benefit from economic activities (see Chapter 3). Resilience and adaptation can therefore be seen as politicised concepts that question societal structures and that can, at their best, address the 'institutional crisis of ... industrial modernity' implied by Beck (1996).

5.4.2.1 Governance theories for understanding adaptation policy

Although the Global Goal on Adaptation prioritises least developed countries, the need for adaptation preparedness is relevant across the globe. For example, in Ireland the National Climate Change Adaptation Framework, 2017, and a range of local planning guidelines and city-specific plans integrate adaptation goals and measures. They are guided by a complex framework of numerous scalar policies, such as the EU's Directive 2007/60/EC on governing flood risk and Ireland's own National Climate Change Strategy, 2007 (Mcdermott & Surminski, 2018; Ray Biswas & Rahman, 2023). Because adaptation governance is not reducible to one clear strategy or policy, there is a need for effective policy integration (Biermann et al., 2020; Cumiskey et al., 2019). This refers to the relationships between different types of actors in different policy areas, such as planning officials, environmental officials and infrastructure maintenance companies. Existing governance mechanisms and norms (rules and priorities in resource allocation for example) can ultimately either hinder or support integration in practice through knowledge transfer, guidance and practical interventions (Cumiskey et al., 2019).

McDermott and Surminski (2018) emphasise the need for an ongoing process of engagement, involvement and participation with local stakeholders in flood risk assessment. Adaptation measures such as flood defences and insurance have to respond to and weigh the trade-offs and potential synergies between flood risk mitigation measures and other needs and priorities at the local scale. Assessments should ideally summon both scientific evidence of risks and experiential knowledge of how risks are perceived and impact everyday life at the local scale. Legitimacy among local stakeholders is crucial, because adaptation to flood risk hinges on community mobilisation. Households and businesses need to be prepared to act in response to a rise in the threat of flooding, and the legitimacy of policy is therefore paramount. Local risk assessment should include economic assessments of potential damage, clarify financial liability – who exactly should pay (government or private businesses for example) – and call for clear cost-share agreements and mechanisms among different levels of stakeholders and government authorities (Mcdermott & Surminski, 2018; Ray Biswas & Rahman, 2023).

In order to contribute to sustainable development as transformation, adaptation policy needs to be guided by the principles of justice and transformative resilience. All four dimensions of justice (distributive, procedural, recognition and reparation) are important for legitimacy and functional governance output. Moreover, integration of policy sectors and actors through guidance, opportunity and resources of knowledge exchange and engagement is crucial. Ideally, adaptation governance should aim at transformational change affecting the core drivers of climate vulnerability (Pelling, 2012).

5.5 Conclusions

Climate change is the defining issue of our time, exacerbating biodiversity loss, inequality, and systemic risks. Human activities – especially fossil fuel use – are the primary drivers of greenhouse gas (GHG) emissions and global warming. Despite international agreements, global warming has already surpassed 1.5°C and is projected to reach 1.8–2.5°C (Darby et al., 2024). The mitigation efforts realised to date remain insufficient and adaptation is now imperative alongside mitigation. Climate justice refers to the inequality characterising the distribution of emissions and climate change impacts across the globe (Oxfam, 2023). Developed countries are responsible for causing most GHG emissions over time, while remaining less vulnerable to climate change impacts. Developing countries and small island states, on the other hand, face severe risks despite minimal contributions. In this context, climate justice requires restorative measures, including financial transfers and technology sharing by developed countries, and just transition policies at the local scale.

The most significant driver of GHG emissions is the energy sector, which accounts for the largest share of global emissions through electricity and heat generation (IPCC, 2022a). Industry follows closely, contributing through manufacturing processes and heavy industrial activities. Agriculture, forestry, and land-use change also play a critical role, particularly through deforestation and intensive farming practices that release carbon and methane. Transport, especially aviation and road vehicles, adds further pressure, while emissions from buildings – linked to heating, cooling, and construction – complete the picture of human-induced climate change.

Beyond these sectoral contributions, systemic and historical causes are to blame (Clark & Szerszynski, 2020). The industrial revolution marked the beginning of a growth-oriented capitalist economy powered by fossil fuels. This economic model, coupled with 'convenience lifestyles' in developed nations, has entrenched high-carbon consumption patterns. Yet, national emission inventories often overlook consumption-based emissions from imported goods, masking the true carbon footprint of affluent societies.

Institutional structures and global inequalities further exacerbate the problem. Fossil fuel dependency is deeply embedded in economic and political systems, making rapid decarbonisation politically challenging. Colonial-era power relations persist in the form of climate injustice, where those least responsible for emissions – developing countries and small island states – face the gravest risks (Ghosh, 2021). This structural imbalance underscores that climate change is not merely an environmental issue but a profound socio-economic and political crisis (Beck, 1996). Addressing the causes of climate change therefore requires not only technological innovation but also transformative shifts in economic systems, consumption patterns, and governance frameworks.

The Paris Agreement (UNFCCC, 2024) forms the core framework of global climate change governance. It was passed in 2015 and, among other things, contains a pledge to limit global warming to well below 2°C, preferably to 1.5°C. To this end, countries submit Nationally Determined Contributions (NDCs) every 5 years, detailing emission reduction targets and strategies. The so-called ratchet mechanism ensures each NDC is more ambitious than the previous one. Emissions are reported bi-annually using the GHG Protocol. At a national scale, a range of governance instruments are used to fulfil the NDCs. Economic tools such as carbon taxes, subsidies for renewables, and cap-and-trade systems are common. COP26 reached an agreement on a global GHG trading system in 2021. However, the Global Stocktake in 2023 revealed a major implementation gap – most large economies (US, China, EU, India) lag in phasing out fossil fuels (Darby et al., 2024). In the same year, Paris Agreement partners made the historic pledge to begin a transition away from fossil fuels. They also pledged to triple renewable energy capacity by 2030.

The NDCs are a bottom-up governance instrument that many criticise for lacking means of enforcement. This is a significant reason for the slow progress in emission mitigation. Several case studies report national interests and short-term economic priorities undermining global commitments. The global carbon market, once fully operational, should discourage carbon leakage resulting from industries relocating to regions with weaker regulations. But so far, global climate governance remains fragmented, with uneven commitments and a lack of enforcement.

Adaptation to the inevitable effects of climate change is imperative and a priority in poor countries located in particularly vulnerable regions. The Paris Agreement set out the Global Goal on Adaptation in 2015, stipulating the prioritisation of vulnerable communities in adaptation efforts and investment

(UNFCCC, 2024). Ideally, adaptation should go beyond short-term fixes to transform societal structures that perpetuate vulnerability (Pelling, 2012). Effective adaptation requires integration across policy sectors, local stakeholder engagement, and attention to justice.

Critics argue that incremental solutions and reliance on technological fixes in climate mitigation and adaptation legitimise the growth-oriented economic system that underpins the root causes of climate change (Beck, 1996). There is an evident need to look beyond current governance solutions to achieve transformative change, potentially by involving degrowth pathways that address the problematic consumption-intensive lifestyles in developed countries and trouble the root causes of climate change by providing low-carbon opportunities. This could, for example, involve public investment in sustainable infrastructure projects that present employment opportunities. Just transition – ensuring fairness for workers and vulnerable communities during the shift away from fossil fuels – is essential. But climate governance is highly politicised, and achieving transformational change requires political will and flexible, participatory, and justice-oriented interventions that maintain legitimacy and effectiveness.

References

Anderson, K. (2023, March 24). IPCC's conservative nature masks true scale of action needed to avert catastrophic climate change. *The Conversation*. Retrieved September 5, 2025, from http://theconversation.com/ipccs-conservative-nature-masks-true-scale-of-action-needed-to-avert-catastrophic-climate-change-202287

Bailey, I. (2017). Spatializing climate justice: Justice claim making and carbon pricing controversies in Australia. *Annals of the American Association of Geographers*, *107*(5), 1128–1143. https://doi.org/10.1080/24694452.2017.1293497

Barrett, J., Peters, G., Wiedmann, T., Scott, K., Lenzen, M., Roelich, K., & Le Quéré, C. (2013a). *Consumption-based GHG emission accounting: A UK case study*. Informa UK Limited. https://doi.org/10.1080/14693062.2013.788858

Barrett, J., Peters, G., Wiedmann, T., Scott, K., Lenzen, M., Roelich, K., & Le Quéré, C. (2013b). Consumption-based GHG emission accounting: a UK case study. *Climate Policy*, *13*(4), 451–470. https://doi.org/10.1080/14693062.2013.788858

Barrett, S. (2012). *The necessity of a multiscalar analysis of climate justice*. SAGE Publications. https://doi.org/10.1177/0309132512448270

Beck, U. (1992). *Risk society: Towards a new modernity*. Sage.

Beck, U. (1996). World risk society as cosmopolitan society: Ecological questions in a framework of manufactured uncertainties. *Theory Culture & Society*, *13*(4), 1–32.

Biermann, F., Van Driel, M., Vijge, M., & Peek, T. (2020). Governance fragmentation. In F. Biermann & R. Kim (Eds.), *Architectures of earth system governance: Institutional complexity and structural transformation*. Cambridge University Press.

Biermann, F., & Kim, R. (2020). Architectures of earth system governance: Setting the stage. In F. Biermann & R. Kim (Eds.), *Architectures of earth system governance: Institutional complexity and structural transformation* (pp. 1–34). Cambridge University Press.

Bomberg, E. (2025). Climate change politics in the age of trump 2.0. *Political Insight*, *16*(3), 13–15. https://doi.org/10.1177/20419058251376783

Bulkeley, H., & Betsill, M. (2005). Rethinking sustainable cities: Multilevel governance and the 'urban' politics of climate change. *Environmental Politics*, *14*(1), 42–63. https://doi.org/10.1080/0964401042000310178

Clark, N., & Szerszynski, B. (2020). *Planetary social thought: The anthropocene challenge to the social sciences*. Polity Press.

Coolsaet, B. (Ed.). (2020). *Environmental justice: Key issues*. Taylor & Francis Group.

Cumiskey, L., Priest, S., Klijn, F., & Juntti, M. (2019). A framework to assess integration in flood risk management: Implications for governance, policy and practice. *Ecology and Society*, *24*(4). https://doi.org/10.5751/ES-11298-240417

Darby, S. J., Hammond, G. P., & Wu, J. (2024). Briefing: Stocktaking global warming: The outcomes of the 2023 Dubai Climate Summit (COP28). *Proceedings of the Institution of Civil Engineers - Energy*, *177*(5), 193–204. https://doi.org/10.1680/jener.24.00005

Evans, J. P., & Thomas, C. (2024). *Environmental governance* (2nd ed.). Routledge.

Folke, C., Polasky, S., Rockström, J., Galaz, V., Westley, F., Lamont, M., Scheffer, M., Österblom, H., Carpenter, S. R., Chapin, F. S., Seto, K. C., Weber, E. U., Crona, B. I., Daily, G. C., Dasgupta, P., Gaffney, O., Gordon, L. J., Hoff, H., Levin, S. A., … Walker, B. H. (2021). Our future in the anthropocene biosphere. *Ambio*, *50*(4), 834–869. https://doi.org/10.1007/s13280-021-01544-8

Ghosh, A. (2021). *The Nutmeg's Curse parables for a planet in crisis*. University of Chicago Press. https://doi.org/10.7208/chicago/9780226815466

Harris, P. (2022). Climate change: International diplomacy and governance from top to bottom. In P. Harris (Ed.), *Routledge handbook of global environmental politics* (pp. 432–444). Routledge.

Harvey, D. (2006). The right to the city. In R. Scholar (Ed.), *Divided cities: The Oxford Amnesty lectures 2003*. Oxford University Press.

Hooghe, L., & Marks, G. (2004). Contrasting views of multi-level governance. In I. Bache & M. Flinders (Eds.), *Multi-level governance* (pp. 15–30). Oxford University Press.

Huber, M. T. (2009). Energizing historical materialism: Fossil fuels, space and the capitalist mode of production. *Geoforum*, *40*(1), 105–115. https://doi.org/10.1016/j.geoforum.2008.08.004

IEA. (2025). *Global energy review 2025*. IEA Publications. https://iea.blob.core.windows.net/assets/5b169aa1-bc88-4c96-b828-aaa50406ba80/GlobalEnergyReview2025.pdf

IPCC. (2007). *Climate change 2007: Synthesis report. Contribution of working groups I, II and III to the fourth assessment report of the intergovernmental panel on climate change* (Core Writing Team, R. K. Pachauri & A. Reisinger, Eds.). IPCC. https://www.ipcc.ch/report/ar4/syr/

IPCC. (2014). *Climate change 2014: Synthesis report. Contribution of working groups I, II and III to the fifth assessment report of the intergovernmental panel on climate change* (Core Writing Team, R. K. Pachauri & L. A. Meyer, Eds.). IPCC. https://www.ipcc.ch/report/ar5/syr/

IPCC. (2022a). *Climate change 2022 mitigation of climate change working group III contribution to the sixth assessment report of the intergovernmental panel on climate change*. Intergovernmental Panel on Climate Change. https://www.ipcc.ch/report/ar6/wg3/

IPCC. (2022b). *Climate change 2022: Impacts, adaptation, and vulnerability. Contribution of working group II to the sixth assessment report of the intergovernmental panel on climate change*. Cambridge University Press. https://www.ipcc.ch/report/sixth-assessment-report-working-group-ii/

IPCC. (2023). *Climate change 2023: Synthesis report. Contribution of working groups I, II and III to the sixth assessment report of the intergovernmental panel on climate change* [Core Writing Team, H. Lee & J. Romero, Eds.). IPCC. https://www.ipcc.ch/report/ar6/syr/

Jackson, T. (2017). *Prosperity without growth: Foundations for the economy of tomorrow* (2nd ed.). Routledge. https://doi.org/10.4324/9781315677453

Keyßer, L. T., & Lenzen, M. (2021). *1.5 °C degrowth scenarios suggest the need for new mitigation pathways*. Springer Science and Business Media LLC. https://doi.org/10.1038/s41467-021-22884-9

King, A., & Cassidy, L. (2025, February 10). Earth is already shooting through the 1.5°C global warming limit, two major studies show. *The Conversation*. Retrieved September 30, 2025, from http://theconversation.com/earth-is-already-shooting-through-the-1-5-c-global-warming-limit-two-major-studies-show-249133

McCauley, D., & Heffron, R. (2018). Just transition: Integrating climate, energy and environmental justice. *Energy Policy, 119*, 1–7. https://doi.org/10.1016/j.enpol.2018.04.014

Mcdermott, T. K. J., & Surminski, S. (2018). *How normative interpretations of climate risk assessment affect local decision-making: An exploratory study at the city scale in Cork, Ireland*. The Royal Society. https://doi.org/10.1098/rsta.2017.0300

Minas, S. (2022). Market making for the planet: The Paris agreement article 6 decisions and transnational carbon markets. *Transnational Legal Theory*, *13*(2–3), 287–320. https://doi.org/10.1080/20414005.2023.2174690

Morrison, T. H., Adger, W. N., Brown, K., Lemos, M. C., Huitema, D., Phelps, J., Evans, L., Cohen, P., Song, A. M., Turner, R., Quinn, T., & Hughes, T. P. (2019). The black box of power in polycentric environmental governance. *Global Environmental Change*, *57*, 101934. https://doi.org/10.1016/j.gloenvcha.2019.101934

Nelson, D. R., Adger, N. W., & Brown, K. (2007). Adaptation to environmental change: Contributions of a resilience framework. *Annual Review of Environment and Resources*, *32*, 395–419. https://doi.org/10.1146/annurev.energy.32.051807.090348

Oxfam. (2023). *Climate equality: A planet for the 99%*. Oxfam GB. https://doi.org/10.21201/2023.000001

Pelling, M. (2012). Resilience and transformation. In M. Pelling, D. Navarrete-Manuel & RedcliftM. (Eds.), *Climate Change and the Crisis of Capitalism* . Routledge.

Piketty, T. (2014). *Capital in the Twenty-First Century*. Harvard University Press.

Ray Biswas, R., & Rahman, A. (2023). *Adaptation to climate change: A study on regional climate change adaptation policy and practice framework*. Elsevier BV. https://doi.org/10.1016/j.jenvman.2023.117666

Ritchie, H. (2026). *CO_2 emissions dataset: our sources and methods*. Our World in Data. Retrieved 17/02/2026, from https://archive.ourworldindata.org/20260129-182256/co$_2$-dataset-sources.html

Ritchie, H., & Roser, M. (2024). *CO_2 emissions. How much CO_2 does the world emit? Which countries emit the most?* Our World in Data. Retrieved Mar 12, 2024, from https://ourworldindata.org/co$_2$-emissions

Ritchie, H., Roser, M. & Rosado, P. (2020). *CO? and Greenhouse Gas Emissions*. Our World in Data. Retrieved Oct 28, 2022, from https://ourworldindata.org/co$_2$-emissions

Sell, S. K. (2018). Who governs the globe. In T. Weiss, & R. Wilkinson (Eds.), *International Organization and Global Governance* (pp. 77–90). Taylor & Francis Group.

Steffen, W., Richardson, K., Johan Rockström, Cornell, S. E., Fetzer, I., Bennett, E. M., Biggs, R., Carpenter, S. R., Wim, d. V., Wit, C. A. d., Folke, C., Gerten, D.,

(FAO, 2024). Mild food insecurity refers to uncertainty about the ability to access food. Moderate food insecurity is a state where a person has to compromise on the variety and quality of food, reduce quantity and occasionally skip meals. Severe food insecurity in turn refers to a state where a person regularly has to go without food for a day or more. Figure 6.1 shows the prevalence of moderate and severe food insecurity based on the UN's Food and Agriculture Organisation (FAO) data from 2024. Food insecurity is most prevalent in Sub-Saharan Africa but remains a concern in mid- and high-income countries, evidencing the rising inequality across the world, even within rich countries. For example, in the UK, the number of people living in food insecure households in 2023 was 7.2 million (Francis-Devine, 2024). In 2023, 17% of children in the UK lived in households experiencing some level of food insecurity, and households from ethnic minority groups are much more likely to experience food insecurity (Francis-Devine, 2024).

While the above definition of food insecurity implicates pillars 1 and 2 (availability and access), food insecurity can result from the indirect impacts of problems in any of the other six pillars outlined in the previous section. Problems with pillars 1, 5 and 6 (availability, sustainability and stability) contribute to food insecurity as a collective risk implicating large groups of people, for example, from national to city scale (Beck, 1992), and pillars 2–4 (access, utilisation and agency) underpin unequal risk

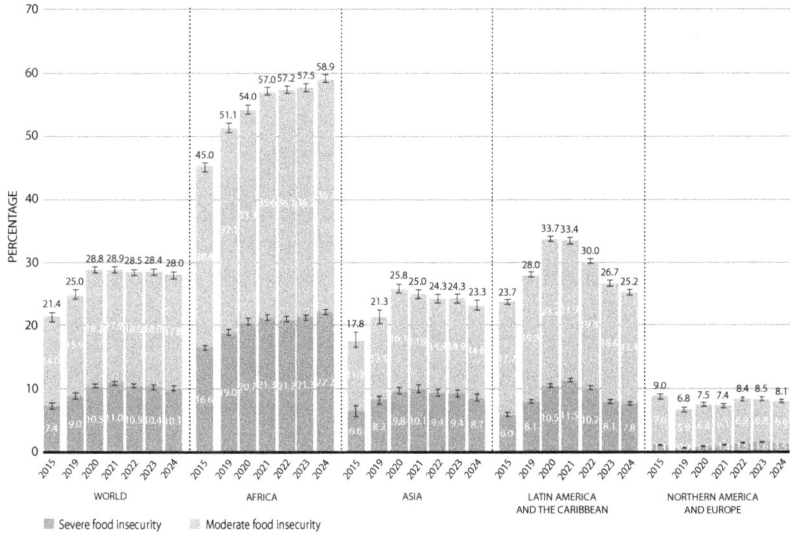

Figure 6.1 The prevalence of food insecurity globally across regions over the past 10 years to 2025, demonstrating the impact of the Covid-19 pandemic in 2020. Oceania is not displayed due to lack of sufficient data.

Data source FAOSTAT: Suite of Food Security Indicators. https://www.fao.org/faostat/en/#data/FS. Adapted from (FAO et al., 2025).

positions across households depending on factors such as education, income and job security.

The price of food is a significant factor in pillar 2, access to food. Centralised commodity markets for soft commodities such as rice and wheat, as well as the food retail sector, are immensely powerful in influencing food prices and can directly influence the prevalence of food insecurity. With climate change, weather-related crop losses can cause fluctuations in the ability to produce food, which can bring prices up. Agency (pillar 4), the ability of people to represent their interests and make choices ensuring access to the means of food production and desired types of food, can easily become compromised in an industrialised food system that prioritises large-scale production and retail units (Jackson et al., 2021). The notion of 'food deserts' refers to mostly urban contexts where there is a lack of retail outlets providing fresh foods to enable healthy diets for local residents (Ron Finley Project, 2022).

Hunger and undernourishment result from prolonged severe food insecurity. FAO measures the prevalence of undernourishment (PoU) as an indicator of hunger at the global and regional level. Undernourishment is estimated on the basis of a country's data on food availability, food consumption and energy needs, and refers to the adequacy of a population's dietary energy intake (FAO, 2024). The regions most affected by hunger are Western Asia, the Caribbean and all subregions of Africa.

Globally, those living in rural areas are more vulnerable to and more severely affected by hunger than urban dwellers. Women tend to be exposed to hunger more often than men, and children are particularly vulnerable as prolonged undernourishment leads to stunted growth and lifelong health impacts. A decade-long global downward trend in hunger (PoU) was broken by the COVID-19 pandemic in 2020 (Figure 6.1). While the proportion of undernourished people is now gradually declining at just below 9% of the global population in 2024 (FAO et al., 2025), at the time of writing in 2025, the proportion of people living with hunger is still higher than in the pre-pandemic years. The war in Ukraine and its impact on the price of energy and food staples pose a further challenge to access to food globally (Economist Intelligence Unit, 2024).

Globally, the steady progress of urbanisation poses both challenges and opportunities for food security (FAO et al., 2023). While the rural poor flock to cities, they become exposed to the same challenges of procuring food as the vast urban poor. But as livelihoods change, so do diets, as urban populations consume more processed and manufactured foods. This arguably creates opportunities for expansion and adding value in the food system, with attendant opportunities for jobs and incomes. Therefore, moving from a small rural subsistence farmer to a worker in urban food manufacturing and retail may provide a rural migrant with a higher and steadier income and better ability to meet household nutrition needs. But ensuring that these opportunities are materialised requires the attention of policymakers and governments to support and encourage the emergence of small and medium-sized agri-food enterprises that can serve the new and expanding urban consumer base (FAO et al., 2023).

A dietary change towards more processed foods often has adverse health impacts. Manufactured foodstuffs have higher fat and sugar content and are linked to obesity and diet-related health conditions. Therefore, urbanisation does not so much mitigate as modify the challenge to food security with a shift of focus from undernourishment to obesity. Obesity rates are rising globally, and overweight and obesity are key underlying factors in rising rates of non-communicable diseases such as cardiovascular diseases and diabetes (FAO et al., 2023). Childhood obesity is on the rise, particularly in middle-income countries, and the third pillar of food security – utilisation – is key to addressing it. The FAO (2023) calls for education and support for changes in food and mobility habits. But the role of the food system, and particularly the agri-food industry, in countering obesity is significant (Jackson, 2025). Many argue that governments must use taxation and regulate advertising to manufactured foods rich in sugar, salt, and fats, making them less attractive to both food businesses and consumers.

6.2.2 The international food system and inequality

Worryingly, Marsden et al. (2019) describe a growing vulnerability, of both food security and the environment, associated with the international food system. From the latter decades of the 20th century, the food system has become highly industrialised. A drive for increased productivity and efficiency of resource use has driven mechanisation, increased reliance on chemical inputs and an increase in average scale of operations across production, manufacturing and retail. In the increasingly globalised food system, production and distribution are becoming concentrated in the hands of large corporate producers and distributors, who remain competitive in the context of international price competition. The rising prices of inputs such as chemical fertilisers and energy have led many smaller farm businesses to abandon the sector altogether or to switch to non-food production in the hopes of better incomes. Efficiency requirements and mechanisation also mean that the agri-food sector now employs fewer people, although the value of its economic contribution to national GDP has grown (Douwe Van Der Ploeg, 2013; Marsden et al., 2019). As a result, the production of food is concentrated in fewer hands at a time when the impacts of climate change and global political unrest are causing further uncertainty for both production and availability.

The proponents of this 'productivist' turn in the food system argue that it is a necessary condition for feeding the growing global population. In order to fulfil the nutritional needs of the over eight billion people populating the planet, food production and manufacturing are best placed in the hands of economically viable businesses that are able to embrace the latest technological developments and the benefits of scale in production. However, critics argue that the structural transformation towards bigger units has endangered food security in the developing country context, where the prevalent form of farming is small-scale subsistence farms of less than 2 ha (Ritchie, 2021).

In a context of weak governance institutions defending access to land, small farmers are not only struggling with the cost-price squeeze and losing market access to larger industrialised producers, but they are also at high risk of displacement and lack options for alternative livelihoods. Subsistence farms often struggle to defend access to land when faced with pressure from large-scale corporate producers, whose interest in farmland as an investment asset has emerged as a significant driver of so-called financialisation in the food system (Marsden et al., 2019). In places, this type of 'land grabbing' by corprate interests has acted as a driver of food insecurity where new large production units have not provided for the needs of local food security previously dependent on local small-scale farming (HLPE, 2020; Mcgreevy et al., 2022).

Alongside a scalar concentration, an internationalisation of food production and distribution networks is also evident (Morgan et al., 2008). Up to mid-20th century, food purchased by households was likely to have been produced by domestic farmers and processed and distributed by domestic companies (Marsden et al., 2019). But thanks to trade liberalisation and effective transport connections, food is increasingly produced in areas of the world where it can be done most cost efficiently, then transported to a third country where there is capacity to process it and finally, to the country where it is packaged and consumed. This means that food products travel long distances, causing emissions and requiring a long shelf life. This requires more external inputs such as fuel for transport and chemical treatments and product development for durability (Behrens et al., 2020).

While the internationalisation of the food system means that we have a broader selection of products to buy in local supermarkets, catering to diverse preferences, consumers are less aware of how and where food has been produced and whether this has been through sustainable practices or not. A severance of sorts is happening between food and its origin, consumers and producers, and production capacity and environmental and climatic factors (Jackson et al., 2021; Morgan et al., 2008). Moreover, relying on far-off places for food production renders availability and access subject to a much broader range of risks, such as supply chain problems and climate change impacts that may differ markedly from those felt in the destination where the food is consumed (Marsden et al., 2019). This underpins a weakening of the agency of consumers and small-scale producers in the food system, while the interests of large-scale producers, traders, and retailers influence the choice of foods available and their production methods and farm-gate and retail pricing.

These challenges demonstrate that in the context of the international food system, food security is a complex topic requiring decisive governance interventions. They also suggest that the industrialised food system in its current form is not delivering on the global commitment to the right to food, which emphasises access to safe and nutritious food for all.

6.2.3 Climate change and environmental sustainability of agricultural production

Climate change compromises food security by limiting productive capacity both on land and in oceans (IPCC, 2023). Although some high latitude regions have experienced positive impacts related to longer and warmer production seasons, climate change has significantly decreased agricultural productivity gains. Extreme weather events such as storms and floods have exposed millions of people to acute food insecurity by destroying crops. Increasing drought poses challenges to crop productivity, particularly in arid and semi-arid regions. Highest water stress is felt in areas that also display highest degrees of food insecurity, in Africa, Asia, Central and South America, Small Islands and the Arctic, with the most severe impact for Least Developed Countries, small-scale food producers, low-income households and Indigenous peoples (IPCC, 2023). Adverse impacts are projected to increase as the world exceeded 1.5°C warming in 2024 and climate change-related destruction of settlements and infrastructure will cause further problems. Therefore, climate change mitigation and adaptation (Chapter 5) are crucial for food security.

But the global food system itself constitutes a significant driver of climate change. In 2022, 22% of the Greenhouse Gas (GHG) emissions driving climate change resulted from agriculture (IPCC, 2022). The share of the whole food system of human-caused climate change emissions is estimated at 26% (Ritchie & Roser, 2024). Mechanisation – tillage and harvest machinery in particular – relies on fossil fuels for energy and contributes significantly to GHG emissions. Agriculture is also a significant driver of land system change and deforestation which limits nature's capacity to absorb CO_2. As diets change to incorporate more animal products such as meat and dairy, the need for more farmlands grows. Animal feed production constitutes an inefficient land use in terms of the provision of calories for human consumption, and meat production, particularly from ruminants such as cows and sheep, emits methane (MH4), a powerful GHG (Behrens et al., 2020).

The impact of industrial agriculture on nature is not limited to driving climate change (Clark & Tilman, 2017). As discussed in Box 2.1 in Chapter 2, land conversion for the cultivation of a narrow range of plant and animal varieties (either with or without the use of genetic modification technologies) disrupts biodiversity and the integrity of the biosphere overall (Dasgupta, 2021). Increased reliance on manufactured inputs such as chemical pesticides and fertilisers has increased leaching and run-off into water courses affecting water quality (Behrens et al., 2020). The expansion of irrigation agriculture in arid and semi-arid regions means that agriculture may account for up to 90% of water use locally, affecting water availability for ecological and other human purposes. While the productivist approach (Wilson & Burton, 2015) focusing on the extraction of maximum possible material and monetary yield from land and labour has increased the quantity of food produced on the planet, it is associated with numerous catastrophic impacts on the underpinning resources that agriculture itself, as well as other life, depends on.

Fortunately, the potential for mitigation of emissions in the land-based sector is also significant. The climate change impacts of agriculture result mostly from livestock and fish production and therefore, a change towards diets that rely more on plant-based foods and integrate less meat and dairy is an effective means of decreasing the climate impact of food (Clark & Tilman, 2017). Cutting livestock production (particularly ruminants) and changes to soil and manure management methods to mitigate methane (CH_4) emissions is essential. Cutting food waste addresses both the need for higher productivity and emissions resulting from the breakdown of organic food waste (IPCC, 2022).

Clark and Tilman (2017) suggest that much of the adverse impacts of modern farming methods can be mitigated through input efficiency – sustainable productivism – such as precision farming, conservation tillage and cover cropping with nitrogen-fixing crops. Precision feeding of livestock supplemented by enzymes and the use of agricultural by-products and waste as animal feed could significantly lower the environmental impact of food production. But to date, the take-up of scientific innovations in agriculture has mainly been motivated by increases in production rather than environmental benefits. The following section explores the notion of sustainability in the food system and governance initiatives for the transformation of the intensive global food system for sustainable development.

6.3 Sustainability in the food system

What then, is a sustainable food system, that will address the structural imbalance, lack of access to production factors and pricing mechanisms underpinning poverty and hunger and the environmental and climate impacts of the globalised industrial food system described above? It is evident that a food-system transformation is required but changes need to be just and well managed so as not to further compromise food security and lead to unequal burdens and impacts.

6.3.1 Sustainable agriculture

Farmers have a range of methods for mitigating the climate and environmental impact of agriculture at their disposal. Using natural pest-control methods and crop rotation to minimise chemical inputs, employing low-impact tillage methods and leaving buffer zones to stop run-off and provide scope for biodiversity are small adjustments for sustainability without significant impact on productivity. 'Regenerative farming' favours methods that align with the local ecosystems' natural regenerative capacities (IPCC, 2023; Mcgreevy et al., 2022). This refers to better alignment between seasons and farming cycles but mainly focuses on methods that support soil formation and local biodiversity, for example through agro-ecological practices. Agro-ecological farming mimics nature's diversity in that it involves producing more than one crop on a single plot of land, for example by combining orchard production with cereal crops and beehives. In this way, the detrimental impact on natural biodiversity is lessened and farming can rely on natural pollinators. Organic farming involves a reduction in chemical inputs such as artificial fertilisers and nutrients, and crop rotation, which helps reduce reliance on such chemicals and engages soil's own regenerative capacities (Behrens et al., 2020). Organic production standards also stipulate better conditions for farm animals, involving free-range chickens and cattle rearing that is associated with a lower need for medical treatments such as the use of antibiotics. Mixed farming, where livestock and crops are produced together, enables a further reduction of chemical inputs as manure can be 'recycled' as a fertiliser.

However, some argue that just increasing the efficiency of productivist farming methods is the best way forward. For example, Clark and Tilman (2017) found that the environmental benefits of organic farming practices are negligible in comparison to highly efficient mechanised farming, mainly due to the difficulty of the precision application of nutrients from manure and the need for larger land areas to achieve the same harvest as with mechanised intensive farming methods. But sustainable productivism does not address the manner in which the international food system marginalises small-scale producers and influences food prices and distribution, compromising access to food for the poor (Jackson, 2017; Mcgreevy et al., 2022).

A move to a sustainable and just food system relying on a sustainable production base would need to be accompanied by interventions addressing the concentration of agency and production factors in international production and retail networks. Re-localisation of parts of the international food system is proposed as one option. 'Local food' that is produced within the vicinity of where it is consumed is more sustainable than food produced, manufactured and transported over long distances (Woods, 2011). Some of this pertains to foregoing so-called 'food miles' referring to the distance travelled and the associated CO_{2e} emissions. But it is important to take into consideration the emissions from inputs such as the energy used to heat greenhouses, for example to produce lettuce in the UK in the winter (Edwards-Jones et al., 2009). Edward-Jones et al. (2009) show that the supply-chain CO_2 emissions of Spanish lettuce sold in the UK were lower than lettuce produced in the UK in the winter. This was mainly because in Spain, greenhouses do not require heating to enable wintertime production. While access to vegetables is crucial for nutrition and therefore food security, their production locally may not always be the most sustainable option and may require seasonal consideration and variation of source.

That said, local food does address the severance of the production-consumption relationship in the food system, addressing the loss of agency by both consumers and small-scale producers. Relying on local produce enables better informed choices on the part of the consumer, enabling vigilance over production methods and the working conditions of the agricultural workforce and the local environment on the part of both governments and consumers. Overall, local scale food systems generally afford better agency to both consumers and producers, including agricultural workers, as the pressures resulting from international competition are less imperative and vicinity supports better transparency.

6.3.2 The right to food, food sovereignty, and justice

Some argue that a sustainable food system requires a reframing of food altogether. Currently, the global industrial food system frames food as a commodity, the production of which should be optimised in accordance with production costs and price signals from consumer demand (Jackson et al., 2021). Considering food as a human right and therefore a public good could provide governments leverage to counter corporate interests and ensure that food produced in the territory of a nation state serves the food security of its population. The notion of food sovereignty refers to the ability of communities at local and regional scales to control food-related assets such as land and technology and to make decisions on how and what food is produced (Woods, 2011). Food sovereignty safeguards the agency of communities and households in determining how and where the food that they consume is produced. This can be done through enabling local food solutions such as urban agriculture and co-operatively owned production units that connect households with producers and embrace the role of food as a cultural value as well as a necessity for sustenance (Morgan et al., 2008). This approach to food addresses many of the vulnerabilities of the industrial food system described above, such as urban food deserts and access to land (Marsden et al., 2019).

As with climate change mitigation and adaptation, it is important to ascertain that any sustainability reforms do not add to the inequalities inherent in the current global food system. Changes in diets (to incorporate more plant-based foods), agricultural production practices (away from large-scale extractive monocultures towards regenerative models), the distribution and access to the means of production (a redistribution of land and technology and, in many cases, re-skilling of those involved in growing food), and the business structures that prevail in the food system (away from large corporations towards shared ownership and foregoing free trade in favour of more stringent regulation) pose numerous trade-offs and conflicts of interest. The burdens implied by these changes are unevenly distributed within the food system, some pertaining to farmers and farm businesses of varying sizes, some to consumers and consumption habits, and some to the ownership of land and business assets.

Therefore, the distributive dimension of justice that focuses on equity and the ability of organisations, communities and individuals to cope with the burden of change is relevant here (Kaljonen et al., 2021). To safeguard distirbutive justice. procedural justice is important in the food system transformation. Different types of stakeholders need to be able to have a say in decision-making. Small farmers and agricultural workers require particular attention as governmental and business interests have dominated the sector (Kaljonen et al., 2021). More transparent, ethical, cooperative and smaller scale business models can be encouraged through subsidies and participatory governance approaches shaping cultural change and new regulation in the food system (Mcgreevy et al., 2022). Changes to diets evoke highly personal preferences and beliefs and cultural norms requiring attention to plural values and knowledges highlighting the improtane also of recognition justice (Coolsaet & Neron, 2020; Kaljonen et al., 2021). In the industrial food system, the quality of food is reduced to variables that can be represented by economic means, for example, the cost of production or the taste, look or rarity of a product. The recognition of food as imbued with nutritional, cultural and environmental values stemming from production processes requiring skill and care and constituting a livelihood yields a broader understanding of quality that needs to be recognised both in pricing and decision-making (Juntti & Downward, 2017).

6.4 Good governance for food security and a sustainable food system

In line with the rest of the book, this chapter considers governance to mean *the institutions and practices of decision-making and intervention regarding issues of public interest, that are undertaken by formal and informal governance actors, often situated in networked relations of power* (Morrison et al., 2019; Sørensen, 2002). Sustainable development implying transformational change in the food system requires change to the governance of food production (agriculture), food prices and agricultural trade. Poverty reduction, infrastructure provision, education, and crisis management are also relevant. As discussed above, dietary change should be seen as a part of the necessary food system transformation for sustainability and therefore, governing for food security needs also to extend to the governance of consumption patterns and habits. Therefore, consumers and communities of rural, urban and Indigenous denomination should be considered as relevant governance actors alongside non-governmental NGOs, community organisations and businesses, in particular, farmers themselves (Kaljonen et al., 2021).

The right to food is championed as a guiding principle for food system governance globally. It informs several governance principles. For example, it obliges governments to set up interventions to safeguard existing access to adequate food (FIAN, 2020). This encompasses access to production factors and, in particular, the protection of existing land rights through the establishment of necessary legislation and documentation. This is particularly important for small farms, which rarely produce food for international trade, and the people who depend on them, who are very vulnerable to hunger (Ritchie, 2021). The obligation to protect the right to food therefore requires measures by states to ensure that enterprises or individuals do not deprive others of their access to adequate food, for example, by legislating against 'land grabs' and ensuring that smallholders' right to land is robustly institutionalised.

To fulfil the right to food, governments must also take measures proactively to strengthen people's access to viable livelihoods. This latter obligation could be seen to refer, for example, to favourable agricultural policies that support farmers' access to technologies and training to guarantee a level of income. The European Union's Common Agricultural Policy (CAP) can be considered such a governance instrument (or a bundle of instruments), as it incorporates subsidies towards technological improvements and education (Gallent et al., 2015). Finally, under the principle of the right to food, governments must have in place social security policies (such as free school meals during holiday periods) that guarantee that food is always accessible in all parts of the country to all socio-economic and demographic groups (FIAN, 2020).

However, prevalent agricultural and food policies tend to fall short of the principles stemming from the right to food (Jackson et al., 2021). For example, the World Trade Organisation (WTO), which provides a global governance forum for the negotiation of the rules and principles underpinning global trade, champions a neoliberal agenda reflecting the logic of the global industrial food system. The WTO Agreement on Agriculture is shaped by the three pillars of improving market access, reducing export subsidies to zero, and limiting the scope of permissible domestic support for agriculture (Fakhri, 2021). This undermines food sovereignty and underpins the concentration of power in corporate hands. For example, it frames any subsidies aimed at supporting small farmers and local businesses as potentially trade-distorting and hindering the modernisation of production processes (Fakhri, 2021).

The High Level Panel of Experts (HLPE, 2020) outlines four global scale governance shifts that are necessary to address the vulnerabilities created by the international food system. Firstly, as the quantity of food produced globally is sufficient and the vulnerabilities lie mainly in access, utilisation and agency pillars of food security, a shift from quantity of production to the multiple dimensions of 'quality' of production and distribution systems is needed. This refers to 'the ability of citizens to exercise agency to shape food systems in ways that meet their needs and preferences, the sustainability of production systems, the resilience of food production and distribution networks, and consideration of the health and nutritional dimensions of food at all stages from production to consumption' (HLPE, 2020, p. 14). Secondly, food system policy needs to interlink with other sectoral policies such as agricultural policy, health policy and even cultural policy. This kind of need for policy integration is common throughout the environmental policy sector

(Cumiskey et al., 2019; Jordan & Lenschow, 2010). A focus on combating hunger is necessary but other dimensions of malnourishment from obesity to diet-related diseases such as diabetes also require action. This implicates pillar 3 of food security and emphasises access to education and a cultural change towards healthier predominantly plant-based diets (Kaljonen et al., 2021).

Finally, all governance solutions need to be tailored to fit the specific context within which they are intended to modulate behaviour (HLPE, 2020). The transition to a sustainable food system may need to take different pathways in different contexts (Wilson & Burton, 2015) where some farms may opt for a shift towards more traditional organic farming styles while others may favour a shift to so-called sustainable productivism where technological solutions mitigate environmental impact. Depending on the more suitable direction of the transformation, there is a differential need to integrate different types of knowledge, including local and Indigenous knowledge into governance. There is therefore a need for a polycentric, networked configuration of governance in food system transformation (Aligica & Tarko, 2012; Morrison et al., 2019).

The next section explores some existing examples of governing agriculture, trade and consumption for a more sustainable food system.

6.4.1 Governing agriculture

As discussed above, agricultural policies are characterised by tension between principles such as productivism, trade liberalisation, sustainability and relocalisation. These agendas represent competing theorisations of how to guarantee the provision of sufficient food for all globally. The WTO was set up to facilitate trade liberalisation and the operation of market forces by dismantling barriers such as import taxes and domestic quality standards or price-distorting domestic production subsidies. Support for local producers and protectionist policies guaranteeing domestic viability and standards, on the other hand, were seen as important for the recovery of domestic production capacity in post-war Europe and are now cast as necessary safeguards for environmental and food quality.

The European Union's Common Agricultural Policy (CAP) is an example of a carefully grafted agricultural policy which aims to uphold food sovereignty in Europe. The objectives of CAP as formulated in the Treaty of Rome in 1957 were as follows (Winter, 1997, p. 118):

- To increase agricultural production by promoting technical progress and by ensuring the rational development of agricultural production and the optimal utilisation of the factors of production, in particular labour.
- To ensure a fair standard of living for the agricultural community, in particular, by increasing the individual earnings of persons engaged in agriculture.

- To stabilise markets.
- To assure the availability of supplies.
- To ensure that supplies reach consumers at reasonable prices.

In 2025, the CAP entails a range of governance instruments but most notably, direct farm payments which aim to ensure a sufficient living standard for farmers so as to maintain the traditional family-farm structure of European food production that is seen to yield multiple benefits beyond food production in the European countryside. The payments are conditional on compliance with regulatory minimum standards for environmental quality, animal welfare and general good farming practice (Juntti, 2012). The CAP has been broadly criticised as market distorting and blamed for flooding global markets with European farm goods. It is by no means a perfect governance instrument and has been associated with surplus production of dairy products and sugar among other things.

In addition to direct payments, so-called agri-environmental policy (AEP) emerged as a part of the CAP reform agreement. The EU Agri-environmental policy was created as a response to the more specific needs of integrating environmental concerns more firmly into agricultural policy, providing alternative channels for direct support into farming and delivering income support for low-income farm practices considered to be environmentally friendly (Buller et al., 2017). This governance instrument can be characterised as a payment for the production of public goods such as habitats, biodiversity and water quality going beyond regulatory requirements. Under the policy, farmers enter a limited-term contract with the respective government agency, agreeing to take some land out of agricultural production for the purposes of environmental conservation activities. These subsidies provide good scope for engaging farmers' professional expertise and local knowledge in delivering the desired environmental outcomes (Juntti, 2012).

While subsidies are popular means for governments to bolster both food availability and the sustainability of production, more top-down regulatory approaches are employed, for example, to curb the use of chemicals, such as nitrate fertiliser, in vulnerable zones (Izcara Palacios, 1998) Unfortunately, despite sophisticated policy frameworks and generally high governance capacity with strong institutions and transparent processes, the shift towards sustainable production has remained an aspiration rather than a reality in the European context. While availability and access to food remain strong, the lack of sustainability suggests that in the face of multiple pressures discussed above, this may not remain the case for long.

6.4.2 Governing trade, retail, and consumption in the agri-food sector

Agricultural policies such as the CAP address the availability and sustainability pillars of food security (HLPE, 2020), but the governance of trade, retail and consumption pertains to the access and utilisation pillars. The FAO guidance on the implementation of the right to food (FAO, 2004) emphasises the need for attention to democracy, economic development policies, food safety and consumer protection policies, and education and awareness raising. It refers to global governance interventions such as the UN Universal Declaration of Human Rights and the International Covenant on Economic, Social and Cultural Rights. It also points to the UN Convention on the Rights of the Child and the UN Convention on the Elimination of All Forms of Discrimination Against Women as of relevance to food security.

Safeguarding the agency of consumers, farm and retail workers, and small farmers requires transparency in the food system. Global certification schemes such as GLOBALG.A.P., which provides an internationally recognised certification scheme for food safety and the environment, workers' safety and welfare, animal welfare, supply chain traceability, and capacity building, present an option to do this (GLOBALG.A.P., 2025). GLOBALG.A.P. is, however, a commercial service that may be out of reach for smaller producers and retailers. Public schemes such as the EU's organic produce certification or the global Fairtrade Foundation that oversees a certification scheme ensuring that small producers in developing countries get a fair price for the produce they export provide further options. Product labels are used to inform the consumer of responsible practices and thus increase their agency in choosing more just and ethical products.

In terms of environmental sustainability, concepts such as virtual water and water footprint have been used to highlight water use as a key aspect of sustainability in food security. Virtual water refers to the volume of water used to produce traded goods, whereas water footprint refers to the impact of the product on water in terms of water consumed in its production and its impact over its lifetime on water quality (Aldaya et al., 2020). Aldaya et al. (2020) suggest that these concepts have helped decision-makers to set more stringent limitations on water use and informed consumers of the impact of the goods that they are purchasing. The terminology of virtual water and water footprints has enabled the media to pick up the issue of water wastage and quality, which in turn has helped raise awareness among consumers and further a cultural change towards more sustainable consumption values and habits.

Informal governance actors therefore play a key role in governing the trade and consumption patterns of food, forming polycentric governance configurations (Morrison et al., 2019). A further example from the UK context is farmers' markets and box schemes, which enable food producers to connect with consumers directly or online and bypass large retail units that have been encouraged through government policy (such as DEFRA's Food 2030 Strategy) and championed by informal governance actors such as the Soil Association and National Lottery (Kirwan & Maye, 2013). While these solutions remain small in scale and often exclusive to the more affluent consumers, they can and should nevertheless play a key role in a more sustainable food system. Proactive governance can help by, for example, providing a formal definition of local food and ensuring that farmers have access to skills and subsidies that enable them to access such local markets and to maintain pricing that renders local food affordable. Nevertheless, Kirwan and Maye (2013) describe the UK's policy approach to food security as predominantly neoliberal and productivist, in alignment with the global food system.

6.5 Conclusions on good governance for food security

Nearly 9% of the global population still suffer from chronic undernourishment and hunger, and 28% experience some degree of food insecurity (FAO et al., 2025). The rural poor, women, and children are particularly vulnerable, which points to uneven risk positions not only globally but also within countries, localities, and even households. The reasons for this are best understood when considering food security as consisting of the six pillars of availability, access, utilisation, sustainability, agency, and stability (HLPE, 2020). While climate change threatens crop productivity and therefore availability of food, the most significant barriers to food security currently pertain to pillars 2–6. Global productive capacity is currently sufficient to feed the global population, but market forces, speculative purchasing of land and commodities, and the interests of large corporate retailers underpin unequal access to and the agency of households and small farmers. The resulting flow of people to cities increases the numbers of urban poor, and while cities offer more opportunities for work and income, urbanisation has underpinned a change in food utilisation and lifestyles that has transformed the risk of food insecurity to one of overweight and obesity for many. Finally, internationalisation and resulting price competition drive a productivist agenda in agriculture, which has resulted in severe environmental impacts of food production, further compromising sustainability.

Presently, large corporate actors such as supermarkets, agri-tech companies and large-scale producers dominate the food system and both farm-gate and consumer prices (Marsden et al., 2019; Morgan et al., 2008). The aspiration to meet global food security needs through ever higher productivity and free trade means that food often travels long distances before it reaches consumers. This compromises the agency of both small-scale producers who constitute the majority of farmers and consumers worldwide (Ritchie, 2021). Local food and regenerative farming practices constitute promising alternatives to the disembedded international food system, and they are being advanced through governance initiatives such as subsidies and product labelling. There are promising projects driven also by informal governance actors that work hard to increase both access to land and reinstate the connection between consumers and producers for increased transparency and accountability in the food system (Mcgreevy et al., 2022). Governments need to take more responsibility for upholding the right to food through governance interventions that curb land grabbing and protect local land rights and fair food pricing (FIAN, 2020).

'Food sovereignty' highlights the importance of local agency, of consumers and producers, in a transformation to a more sustainable food system. Food sovereignty hinges on principles such as equity, fairness and sustainability as underpinning food security for all at the global scale (FIAN, 2020). But the international food system consisting of a variety of interconnected actors renders governing for sustainable food security a complex exercise. To transform the global food system for sustainable development, governance needs to focus on the environmental sustainability and climate resilience of food production, access to land, the rights of small farmers and agricultural workers, and the ability of people and communities to make informed and sustainable choices when it comes to the food that they consume.

Currently, food system governance is characterised by the tension between two competing agendas. A productivist liberal agenda promotes internationalisation and technological development as a route to food security. It aligns with the current food system proposing enhancements mainly to the sustainability of the production base. The alternative more holistic sustainability agenda aligns with the principle of food sovereignty and focuses on access and agency at the local scale. It promotes solutions that re-connect producers and consumers and prioritise environmental regeneration and local agency. This chapter has argued in favour of this latter agenda. But decisive action from both formal and informal stakeholders is needed if the food system is to adhere to the principles of fairness and sustainability embodied in the notions of food sovereignty and the right to food.

References

Aldaya, M. M., Garrido, A., & Llamas, R. (2020). Water footprint and virtual water trade: The birth and growth of a new research field in Spain. *Water*, *12*. https://doi.org/10.3390/w12092641

Aligica, P. D., & Tarko, V. (2012). Polycentricity: From Polanyi to Ostrom and beyond. *Governance: An International Journal of Policy, Administration, and Institutions*, *25*(2), 237–262.

Beck, U. (1992). *Risk society: Towards a new modernity*. Sage.

Behrens, P., Bosker, T., & Ehrhardt, D. (2020). *Food and sustainability*. Oxford University Press.

Buller, H., Wilson, G., & Holl, A. (2017). *Agri-environmental policy in the European Union*. Routledge.

Canfield, M., Anderson, M. D., & Mcmichael, P. (2021). *UN food systems summit 2021: Dismantling democracy and resetting corporate control of food systems*. Frontiers Media SA. https://doi.org/10.3389/fsufs.2021.661552

Clark, M., & Tilman, D. (2017). *Comparative analysis of environmental impacts of agricultural production systems, agricultural input efficiency, and food choice*. IOP Publishing. https://doi.org/10.1088/1748-9326/aa6cd5

Coolsaet, B., & Neron, P. (2020). Recognition and environmental justice. In B. Coolsaet (Ed.), *Environmental justice: Key issues* (pp. 52–63). Taylos & Francis Group.

Cumiskey, L., Priest, S., Klijn, F., & Juntti, M. (2019). A framework to assess integration in flood risk management: Implications for governance, policy and practice. *Ecology and Society, 24*(4), Article 17. https://doi.org/10.5751/ES-11298-240417

Dasgupta, P. (2021). *The economics of biodiversity: The Dasgupta review abridged version*. HM Treasury.

DEFRA. (2024). *Official statistics United Kingdom food security report 2024: Theme 2: UK food supply sources*. HMG. https://www.gov.uk/government/statistics/united-kingdom-food-security-report-2024/united-kingdom-food-security-report-2024-theme-2-uk-food-supply-sources

Douwe Van Der Ploeg, J. (2013). Peasant-driven agricultural growth and food sovereignty. *Journal of Peasant Studies*, *40*(6), 999–1030.

Economist Intelligence Unit. (2024). *Worldwide cost of living: How soaring inflation has affected prices globally*. The Economist.

Edwards-Jones, G., Plassmann, K., York, E. H., Hounsome, B., Jones, D. L., & Milà i Canals, L. (2009). Vulnerability of exporting nations to the development of a carbon label in the United Kingdom. *Environmental Science & Policy*, *12*(4), 479–490. https://doi.org/10.1016/j.envsci.2008.10.005

Fakhri, M. (2021). *A trade agenda for the right to food*. Springer Science and Business Media LLC. https://doi.org/10.1057/s41301-021-00305-0

FAO. (2004). *Voluntary guidelines to support the progressive realization of the right to adequate food in the context of national food security*. Food and Agriculture Organisation. https://openknowledge.fao.org/server/api/core/bitstreams/307a8e6b-c478-49ba-8a29-f97c825d5770/content

FAO. (2024). *Hunger*. Food and Agriculture Organization of the United Nations. Retrieved June 27, 2024, from http://www.fao.org/hunger/en/

FAO, IFAD, UNICEF, WFP, & WHO. (2023). *The state of food security and nutrition in the world 2023. urbanization, agrifood systems transformation and healthy diets across the rural–urban continuum*. FAO; IFAD; UNICEF; WFP; WHO. https://doi.org/10.4060/cc3017en

FAO, IFAD, UNICEF, & WHO. (2025). *The state of food security and nutrition in the world 2025 – Addressing high food price inflation for food security and nutrition*. https://doi.org/10.4060/cd6008en

FIAN. (2020). *Right to adequate food and nutrition, and to food sovereignty*. FIAN. https://www.fian.org/en/right-to-adequate-food-and-nutrition-and-to-food-sovereignty-3/

Francis-Devine, B. (2024). *Who is experiencing food insecurity in the UK?* UK Parliament, House of Commons Library, Insight. Retrieved April 7, 2024, from https://commonslibrary.parliament.uk/who-is-experiencing-food-insecurity-in-the-uk/

Gallent, N., Hamiduddin, I., Juntti, M., Kidd, S., & Shaw, D. (2015). *Introduction to rural planning: Economies, communities and landscapes* (2nd ed.). Routledge. https://doi.org/10.4324/9781315749280

GLOBALG.A.P. (2025). *About GLOBALG.A.P.* Retrieved January 4, 2026, from https://www.globalgap.org/about/

HLPE. (2020). *Food security and nutrition building a global narrative towards 2030 a report by the high level panel of experts on food security and nutrition of the committee on world food security*. HLPE.

IPCC. (2022). *Climate change 2022 mitigation of climate change working group III contribution to the sixth assessment report of the Intergovernmental Panel on Climate Change*. Intergovernmental Panel on Climate Change. https://www.ipcc.ch/report/ar6/wg3/

IPCC. (2023). *Climate change 2023: Synthesis report. Contribution of working groups I, II and III to the sixth assessment report of the intergovernmental panel on climate change* [Core Writing Team, H. Lee & J. Romero, Eds.). IPCC. https://www.ipcc.ch/report/ar6/syr/

Izcara Palacios, S. (1998). Farmers and the implementation of the EU nitrates directive in Spain. *Sociologia Ruralis*, *38*(2), 144–162. https://doi.org/10.1111/1467-9523.00069

Jackson, P., Guadalupe, M., Ferre, R., Candel, J., Davies, A., Derani, C., De Vries, H., Dragović-Uzelac, V., Hoel, A. H., Holm, L., Mathijs, E., Morone, P., Penker, M., Śpiewak, R., Termeer, K., & Thøgersen, J. (2021). *Food as a commodity, human right or common good Different framings of food may shape food policies and their impact. Despite acknowledging food systems' complexities, the European Commission's Farm to Fork Strategy still addresses food as a commodity instead of a human right or common good. Nature Food*, 2(3), 132–134.

Jackson, T. (2017). *Prosperity without growth: Foundations for the economy of tomorrow* (2nd ed.). Routledge. https://doi.org/10.4324/9781315677453

Jackson, T. (2025). *The care economy*. Polity.

Jordan, A., & Lenschow, A. (2010). Policy paper environmental policy integration: A state of the art review. *Environmental Policy and Governance, 20*(3), 147–158. https://doi.org/10.1002/eet.539

Juntti, M. (2012). Implementing cross compliance for agriculture in the EU: Relational agency, power and action in different socio-material contexts. *Sociologia Ruralis*, *52*(3). https://doi.org/10.1111/j.1467-9523.2012.00564.x

Juntti, M., & Downward, S. D. (2017). Interrogating sustainable productivism: Lessons from the 'Almerían miracle'. *Land Use Policy*, *66*, 1. https://doi.org/10.1016/j.landusepol.2017.04.016

Kaljonen, M., Kortetmäki, T., Tribaldos, T., Huttunen, S., Karttunen, K., Maluf, R. S., Niemi, J., Saarinen, M., Salminen, J., Vaalavuo, M., & Valsta, L. (2021). Justice in transitions: Widening considerations of justice in dietary transition. *Environmental Innovation and Societal Transitions*, *40*, 474. https://doi.org/10.1016/j.eist.2021.10.007

Kirwan, J., & Maye, D. (2013). Food security framings within the UK and the integration of local food systems. *Journal of Rural Studies*, *29*, 91. https://doi.org/10.1016/j.jrurstud.2012.03.002

Marsden, T., Faus, A., & Sonnino, R. (2019). Reproducing vulnerabilities in agri-food systems: Tracing the links between governance, financialization, and vulnerability in Europe post 2007–2008. *Agrarian Change19*: 82–100. https://doi.org/10.1111/joac.12267

Mcgreevy, S. R., Rupprecht, C. D. D., Niles, D., Wiek, A., Carolan, M., Kallis, G., Kantamaturapoj, K., Mangnus, A., Jehlička, P., Taherzadeh, O., Sahakian, M., Chabay, I., Colby, A., Vivero-Pol, J., Chaudhuri, R., Spiegelberg, M., Kobayashi, M., Balázs, B., Tsuchiya, K., ... Tachikawa, M. (2022). *Sustainable agrifood systems for a post-growth world*. Springer Science and Business Media LLC. https://doi.org/10.1038/s41893-022-00933-5

Morgan, K., Marsden, T., & Murdoch, J. (2008). Networks, Conventions, andRegions: Theorizing 'Worlds of Food'. In K. Morgan, T. Marsden & J. Murdoch (Eds.), *Worlds of Food: Place, Power, and Provenance in the Food Chain* (pp. 7–25). Oxford University Press.

Morrison, T. H., Adger, W. N., Brown, K., Lemos, M. C., Huitema, D., Phelps, J., Evans, L., Cohen, P., Song, A. M., Turner, R., Quinn, T., & Hughes, T. P. (2019). The black box of power in polycentric environmental governance. *Global Environmental Change*, *57*, 101934. https://doi.org/10.1016/j.gloenvcha.2019.101934

Ritchie, H. (2021). *Smallholders produce one third of world's food, less than half of what many headlines claim*. Our World In Data. https://ourworldindata.org/smallholder-food-production

Ritchie, H., & Roser, M. (2024, /01/22). *CO_2 emissions. How much CO_2 does the world emit? Which countries emit the most?* Our World in Data. Retrieved March 12, 2024, from https://ourworldindata.org/co$_2$-emissions

Ron Finley Project. (2022). *The Ron Finley Project*. Ron Finley Project. Retrieved March 25, 2022, from https://ronfinley.com/

SAM. (2020). *Towards a sustainable food system. Independent expert report*. European Commission DG for Research and Innovation. https://doi.org/10.2777/37244

Sørensen, E. (2002). Democratic theory and network governance. *Administrative Theory & Praxis*, *24*(4), 693–720. https://doi.org/10.1080/10841806.2002.11029383

Wilson, G. A., & Burton, R. J. F. (2015). 'Neo-productivist' agriculture: Spatio-temporal versus structuralist perspectives. *Journal of Rural Studies*, *38*, 52. https://doi.org/10.1016/j.jrurstud.2015.02.003

Winter, M. (1997). New policies and new skills: Agricultural change and technology transfer. *Sociologia Ruralis*, *37*(3), 363–381.

Woods, M. (2011). *Rural geography III*. SAGE Publications. https://doi.org/10.1177/0309132510393135

Part 3

Lessons for good governance for sustainable development

7 Identifying good governance for sustainable development

Learning objectives

- What are the main challenges for sustainable development as transformational change in the 21st century?
- How can the principles of regeneration, integration, legitimacy and reflexivity benefit governance in overcoming the above challenges?
- How can we evaluate whether governance is delivering on sustainable development?
- What is the purpose of ex-ante and ex-post evaluation in the governance cycle?
- How can evaluation drive change in governance to support transformation rather than incremental change?

7.1 Introduction

The 21st-century multi-crises pose an urgent agenda for sustainale development. The planetary crisis threatens the very basis of our societies and economies (Folke et al., 2021; Richardson et al., 2023), and the inequality crisis manifests in growing disparity in socio-economic circumstances and in vulnerability to the planetary crisis (Chancel et al., 2026; Oxfam, 2023). These two crises underpin much of the current political and social instability visible in the rise of populist leaders, weakening of the core institutions of democracy and destabilisation of the international balance of power (Piketty, 2014; World Bank Group, 2025; WEF, 2025).

Literature points to broadly two alternative paths for sustainable development: to adjust to new and unpredictable reality and mitigate its impacts, or to reset the economic system and our relationship with the Earth towards one where fossil fuel use, deforestation and land system change are significantly lower and efforts to restore key elements of the system are rewarded as development. Given that the planet has already reached 1.5°C warming in 2024 (King & Cassidy, 2025), the likely reality is going to be a combination of the above two paths. But sustainable development also stipulates that any ostensible development trajectory must advance in a fair and predictable manner

DOI: 10.4324/9781003179207-10

that sustains people, flora, fauna and indeed the whole planet within a managed transformation. This necessitates a fundamental change to the prevalent logic of societal decision-making and, indeed, governance.

Chapter 3 set out the definition of sustainable development as transformational change extending to the core of our societal systems, such as the global market economy and the international food system, as well as to the dominant social norms that tend to prioritise monetisable value. Transformational sustainability integrates a planetary perspective, a regenerative care-oriented economic order, and a fine-tuned understanding of justice as a guiding principle of decision-making. Chapter 4 explored different theoretical lenses that support critical assessment of governance and point to approaches that are able to integrate the requirements of transformational sustainability. It argued that good governance is governance that is effective – legitimate, addresses the root causes of unsustainability and achieves fair and sustainable outcomes. Part 2 of the book explored the two issues of climate change and food security central to sustainable development, and the governance approaches and instruments available to address them. It identified implementation gaps and their underpinning mechanisms such as the economic growth imperative and system complexity. Considering the challenges of the 21st-century governance context (Chapter 2) and the complexity of the required sustainability transformation (Chapter 3), there is no easy recipe for good governance for sustainable development. Perhaps the most important requirement is for governance to embrace complexity. Governance needs to be informed by systems thinking, recognising that our everyday lives are shaped by systems of technologies, social practices, business models, regulations and societal norms within which behaviours and practices are premised (Voulvoulis et al., 2022). Planetary justice (Gupta et al., 2023) can provide normative steer.

In line with earlier chapters, governance is here taken to mean *the institutions and practices of decision-making and intervention regarding issues of public interest, that is undertaken by formal and informal governance actors, often situated in networked relations of power* (Morrison et al., 2019; Sørensen, 2002). The followign two sections consider the main challenges that governing for sustainable development needs to overcome and propose four priniples that, in this context, denote good governance. The latter half of the chapter explores the role of policy evaluation in assessing both how governance is undertaken and whether its aims align with transformative sustainable development.

7.2 Challenges and principles for good governance for sustainable development

If we understand sustainable development as a societal *transformation* towards *equitable and regenerative* relations among people and the planet, governance for sustainable development needs to address the driving forces of unsustainability. This involves challenging some dominant norms and institutions and addressing difficult trade-offs, often between conventional indicators of development such as economic growth and the limits set by the planetary boudnaries (Steffen et al., 2015). The below sections first set out the key challenges for governance for sustainable development and subsequently explore a range of principles that can guide governance in the difficult task of achieving transformational change.

7.2.1 The challenges for sustainable development as transformational change

7.2.1.1 The growth imperative

Currently, the global capitalist market economy acts as a hegemonic organising force of society. Its driving force is economic growth, which, through the operation of global free markets, is assumed to increase the overall capacity for human needs-fulfilment. To fit this dominant order, sustainable development needs to be delivered through solutions that can be plugged into the economy in a profitable manner (Jackson, 2017; Lélé, 1991; O'Neill, 2017). Interventions, for example for climate change mitigation, that do not align with the economic growth rationale become framed as trade-offs for the ostensible benefits of economic growth. This juxtaposition frequently stalls progress towards sustainable solutions both in terms of governance and practice and perpetuates unabated emissions and resource extraction. While commodifiable sustainability solutions that align better with the gorwth-imperative, such as carbon-neutral technologies, can progress some aspects of sustainability, they often cause new problems and further inequality.

Bailey (2017) for example, illustrates how short-term economic prospects hold enormous sway in both the public and political domains at national scale (Chapter 5). In Australia, the potential impact of carbon pricing on small and medium-sized businesses, particularly agriculture, toppled the Clean Energy Mechanism (CEM) that aimed to set a price on GHG emissions to incentivise their mitigation (Bailey, 2017). While there is no doubt that over 1.5°C warming will have a cost to the economy in terms of heightened occurrence of phenomena like floods, droughts, and fires that degrade food security and damage infrastructure, these long-term costs remain mostly unacknowledged in decision-making. As long as agriculture operates in the context of the global market economy where economic viability is decisive for survival, emission pricing alone is unlikely to gain strong legitimacy in the sector. A more fundamental change to the rationale of the economy is needed.

On the other hand, the need to foster economic growth has inspired commercial emission mitigation solutions, often so-called clean technologies that can first be incentivised through a subsidy with the idea that once markets reach a tipping point, they will become economically viable. However, these innovations tend to harness natural resources for further extraction, merely displacing harms and creating new scarcities. For example, the production of carbon-neutral biofuels to replace fossil-based ones occupies agricultural land, potentially taking it out of food production (McDonagh, 2014).

Biofuels are often produced in territories that are vulnerable to land grabs and extractive production practices that cause environmental harms and disadvantage local populations, at worst, compromising local food security (Koizumi, 2015). Competition in the biofuel market means that their production is rarely possible for local small-scale farmers and falls in the hands of multinational corporations and investors. The example of biofuels, therefore, also demonstrates how profit-oriented sustainability solutions drive inequality. A subsidy for biofuel production risks a range of unexpected outcomes and perpetuates an extractive approach to planetary resources in a context where growing energy consumption is necessary to feed economic growth.

Finally, even public-goods-based sustainability solutions such as ecosystem services become commodified in the context of the prevalent profit-oriented planning and land-use regime (Brenner & Elden, 2009; Kronenberg, 2025). Nature-Based Solutions such as natural drainage solutions consisting of increased permeable surfaces and plant varieties that thrive in changeable hydrological conditions are framed as win-win solutions, both serving urban resilience in terms of flood risk mitigation and contributing real estate value by providing improved 'liveability'. These monetisable services and benefits then become prioritised in urban planning at a cost to the broader community value of greenspaces that could otherwise warrant more diverse types of uses and users (human or otherwise) of the spaces in question (Juntti et al., 2025). Commodification narrows the range of acceptable uses of public goods, excluding diverse needs and values.

7.2.1.2 Complex socio-technical systems

Complexity, of both governance issues such as climate change and food security, and the 21st-century crises context, poses a key challenge for governance for sustainable development. The theory of the network society captures some of this compelxity (Chapter 2) as does the notion of complex systems as organising structures of society (Castells, 2004; Voulvoulis et al., 2022). Both of these phenomena can be see as driven by the global economic system and its implicit gowth impreative. The neoliberal turn of the 1990s centred principles such as economic efficiency and the free flow of goods and services in the global economy (Weiss & Wilkinson, 2018). This furthered the internationalisation of supply networks and provided further impetus for the adoption of technological solutions to support the creation of economies of scale. But the resulting interconnectedness of localities and stakeholders in complex commodity networks now appears as a key driver of further inequality and renders governance prone to unexpected outcomes.

Voulvoulis et al. (2022) for example, describe the unexpected side-effects of environmental policies that overlook the complexity of the contexts within which the intended behavioural changes should take place. In the US, a policy aimed at curbing habitat loss from land-use change as a result of hydro-power expansion resulted in increased reliance on fossil energy and

GHG emissions. Whereas in the UK efforts to curb CO_2 emissions inadvertently led to growth in diesel engine vehicles and attendant air pollution. The global food system is an example of one such complex socio-technical network that conditions availability and access to food and limits the scope for sustainability in food production and consumption. Chapter 6 describes how increasing internationalisation and the growing role of technology in agriculture and food manufacturing have driven detrimental environmental impact of the agri-food sector, led to a structural change towards larger farm sizes and a shift of ownership to large corporate producers, decreasing the agency of consumers and small farmers (Marsden et al., 2019; McDonagh, 2014; Wilson & Burton, 2015). Crucially, food insecurity continues to affect a significant minority (approximately 30%) of the world's population, with issues concerning both availability and access to food the most prevalent causes (FAO et al., 2023).

7.2.1.3 The gaps in governance ambition and implementation

In the case of key sustainable development interventions, such as climate mitigation policy and the UN SDGs, set targets often either fall short of the required changes or fail to achieve the projected change in practice (Biermann et al., 2020; Darby et al., 2024). According to Biermann et al (2022), by 2022, the SDGs had mainly achieved changes at the level of political rhetoric. Critics suggest that the Paris targets for mitigating GHG emissions remain woefully unambitious (Anderson, 2023). Anderson (2023) claims that the most polluting economies need to commit to reducing their consumption-related emissions much more quickly, achieving carbon neutrality by 2035 instead of 2050, to avert significant warming. Moreover, the Nationally Determined Contributions (NDCs) are falling short even of the insufficient Paris commitments (UNFCCC, 2024). The global GHG accounting and reporting system suggest that a significant emissions gap exists, whereby the NDCs submitted in 2022 leave the world on a path to 2.1–2.8°C warming (IPCC, 2023).

Despite the mounting scientific evidence signalling an urgent need to act, the amenability of the global political environment to climate mitigation has fluctuated over the years. The economic growth imperative and system complexity account for much of this lack of progress and ambition. For example, the re-election of Donald Trump as president in 2024 saw the US, the biggest global GHG emitter, drop several global commitments, including the Paris Agreement, and begin a renewed campaign of acquiring and accessing new sources of oil.

In the absence of top-level political will, stabilising the Earth system and the climate depends on a cultural change and innovative push from non-state stakeholders, including communities and organisations at sub-national scales. But, limited agency at local scale adds to the lack of progress towards targets. Substantial change to the consumption-oriented lifestyles in developed countries, including reduction in energy consumption, is more difficult in the context of the consumption-powered economic system and runs against the interests of its powerful beneficiaries. The Australian example (above and Chapter 5) demonstrates how the imperatives of growth and competitiveness set by the current economic system limit ambitious mitigation policies on the ground.

7.2.1.4 Unequal risk positions and environmental injustices

'Climate justice' draws attention to historic and current inequalities in terms of responsibility for emissions, risk burdens from their impacts and ability to adapt to already inevitable changes caused by climate change (Barrett, 2012; Coolsaet, 2020). The challenge that climate justice poses for governance focuses on holding the biggest emitters accountable while simultaneously ensuring that those who need to adapt are able to do so while meeting an acceptable standard of social needs for all (Raworth, 2017). This question implicates all scales of governance from the global to the very local and poses a substantial challenge to both adaptation efforts and trajectories towards net zero.

While the UNFCCC and the Paris Agreement have some mechanisms to recognise the unequal risk positions of their different signatories, recent COPs have shown that actual delivery on the pledged transfers of finance and technology from those who have, to those who want, rarely meet the set targets (Darby et al., 2024). Moreover, the contentious discussions around climate reparations demonstrate the difficulty of articulating, let alone measuring, the impact of centuries of resource extraction and subordination during the colonial era, industrialisation and, more recently, unequal international production and trade networks (Ghosh, 2021; Oxfam, 2023; Sze, 2020). Meanwhile, the profit imperative and complex systems themselves continue to accelerate inequality (see above and Chapters 2 and 6).

7.2.1.5 The crises of knowledge and legitimacy

Finally, the crisis of knowledge (Chapter 2) challenges efforts to gain legitimacy in governance and manifests in the rise of so-called populism, the cynical use of mis- and disinformation for political gains. The complexity of the 21st-century crises and their drivers has been seized by populist political actors as an opportunity to deliberately obscure the public's vision to gain influence. Public denouncements of expert knowledge by politicians (Sze, 2020) feed into phenomena such as climate scepticism, which creates confusion and further undermines political will to adopt mitigation and adaptation measures that may compromise short-term economic ends (Ray Biswas & Rahman, 2023).

Beyond this, there is an emerging acknowledgement that scientific knowledge alone will not lead to legitimate solutions to the planetary crises. Despite abundant scientific evidence and efforts dating back decades, the world has now crossed at least six of the nine risk boundaries of the Earth system and surpassed the maximum 1.5°C warming target (Richardson et al., 2023). As discussed above, many of the commercial solutions have instead created new scarcities and inequalities. There are therefore growing calls for non-Western ways of knowing about the planet, the position of humans in relation to nature and about the human experience of risks and needs (Clark & Szerszynski, 2020; Ghosh, 2021; Sze, 2020).

In the 21st century, good governance needs to both facilitate and be guided by the emergence of alternative futures beyond those afforded by economic growth and technological innovations that serve it. For this purpose, scientific ways of knowing need to be accompanied by decolonial, alternative perspectives and knowledges that acknowledge people's experienced realities and are capable of supporting the development and realisation of a radically different relationship with the planet.

7.2.2 Governance principles for transformation in the 21st century

This section sets out four principles for good governance grounded in an understanding of sustainable development as transformational change. It draws from the theorisations of governance and power covered in this book and aims to respond to the above challenges.

7.2.2.1 Regenerative approach to the Earth system

Many scholars stipulate a scalar shift in the scope of governance from the global to the planetary as necessary for sustainable development. This is because, in the context of the Anthropocene, the reach and influence of human actions, the pressing governance issues of the 21st century and indeed the crises which we are trying to manage through governance, extend to the planetary scale (Brenner & Elden, 2009; Clark & Szerszynski, 2020; Folke et al., 2021; Gupta et al., 2023; Infante-Amate et al., 2024; San Martín & Wood, 2021; Steffen et al., 2015).

This shift in the imaginary of governance is exemplified by the notion of Earth system boundaries (Steffen et al., 2015). The term Earth system encompasses not just human societies and economies but those natural systems on which we depend, such as the geosphere, biosphere and the atmosphere, drawing non-human actors and non-living features of the planet within the focus of governance (Folke et al., 2021). This expansion of scale to the planetary can be seen as a significant reorientation. The impacts of the planetary crises and the climate crisis in particular signal the fundamental interdependence of humans, non-human species and the non-living features of the Earth system and emphasise the need for maintaining planetary equilibrium by affording each system component the ability to function as required.

The planetary perspective therefore recognises that not only humans, but non-human species and other key components of the planetary system, hold significant agency. This 'decentring' of humans in favour of the 'more-than-human' perspective (Whatmore, 2006) demands that we revisit the value basis of governance, where 'issues of shared interest' encompass not just human interest, but that of the planetary system and humans as a part of it. The notion of planetary justice and its sub-fields such as recognition justice and epistemic justice reflect this shift (San Martín & Wood, 2021). The rights-based approach to the environment offers further help (Chapters 3 and 6). There is therefore a need to think in terms of more-than-human partnerships in governance and assign rights to non-human species and features of the environment such as rivers and forests (Cohen et al., 2023; Mcgreevy et al., 2022).

As the planetary perspective reveals humans as profoundly dependent on the complex and dynamic Earth system, it prescribes a shift from the predominantly extractive approach to the planet's resources and functions towards one focused on restoration, regeneration and care (Gupta et al., 2023; Jackson, 2025; Sze, 2020). Where the ecosystem services approach signified a first step in understanding the value of nature for humans, in terms of the functions and resources that underpin our societies and economies, the planetary perspective underlines our position as an interdependent component of a much larger system, the sustainability of which is profoundly in our self-interest.

7.2.2.2 *Integration between policy domains*

Unintended 'side effects' occur when a governance intervention solves or partially solves a problem it is intended to address, but in the course of it, causes a new one. As discussed above, proliferating complex commodity networks such as the international food system underpin these 'spillover' or 'displacement' effects (Biermann & Kim, 2020). Subsidised biofuels discussed above constitute one example, where combustion engine-based GHG emissions mostly from European and North American countries are abated at a cost to food security in Asia where biofuel production is being encouraged by subsidies (Koizumi, 2015). The problem is that governance systems fail to match the geographical reach of complex systems, appearing disconnected or fragmented, either across scales or sectors of governance (Biermann et al., 2020). Addressing this through coordination of policy aims of relevant governance domains is therefore central to good governance for sustainable development (Weitz et al., 2017).

Governance fragmentation can be assessed along a spectrum ranging from synergistic, through cooperative to conflictive institutional structures, norms and actor constellations (Biermann et al., 2020). It reflects the extent to which two or more policy areas (for example, climate change and food security) have similar aims and norms and can work together in a coordinated manner. Governance integration can remedy fragmentation either by means of structural consolidation through committees and mandated changes to decision-making procedures to render them more open to collaboration, or through functional means relying on governance stakeholders, officials, talking to each other and identifying coinciding agendas, remits and benefits (Cumiskey et al., 2019). Governance, after all, is ultimately communication between people. Weitz et al (2017, p. 168) suggest that the prevalence of shared understandings of governance issues (consistency) and coordination of implementation and evaluation practices and rules are indicators of integration between two policy domains.

The 'food, water, energy' nexus is an example of an integrated focal point for sustainability governance as that is where many of the most pressing governance conflicts and synergies arise. For example, to address the conflicting goals of biofuel and food security policies, a more integrated approach to land-use governance should be adopted. This would then enable the coordinated consideration of several policy challenges ranging from renewable energy production to food security, and both structural and functional integration of these separate policy domains.

Environmental policy integration (EPI) refers to the consideration of the environmental impacts of all policies, regardless of the policy area in question (Jordan & Lenschow, 2010). 'Strong' EPI grants 'principled priority' to environmental objectives while 'balancing' economic, social and environmental concerns, and seems particularly pertinent in the light of the emissions gap plaguing progress in climate mitigation and adaptation (Weitz et al., 2017).

Scalar integration is also important. Where international agreements usually stipulate a certain degree of freedom of interpretation, enabling nation states and local and community organisations to adapt agreements and rules to national and local circumstances, this must not lead to the watering down of governance goals enabling further spillover (Weitz et al., 2017). Nevertheless, this is where a fine balance must be achieved between the coherence of governance goals and outcomes and the agency of local communities and knowledges. In line with reflexive governance principles (Voß & Kemp, 2006, Brousseau et al., 2012) also the EPI literature places emphasis on trust, ownership, and learning among formal and informal governance actors and across policy domains (Weitz et al. 2017: 170).

7.2.2.3 Legitimacy and justice

Legitimacy (of a policy or a stakeholder) is seen to stem from roughly three sources: recognition by or at least alignment with law, popular acceptance and alignment with dominant norms (Bulkeley, 2012). Bulkley (2012: 2429) provides a relational definition of legitimacy which is realised when 'actors within a given community come to regard institutions, rules, and norms as those with which they ought to comply because they are right, proper, or appropriate'. As discussed in Chapter 2, a crisis refers to a situation where societal norms and systems, possibly including legal frameworks, are rendered inadequate by developments that challenge the status quo and call for transformational change (Kuipers, 2006). In this context, the relational understanding of legitimacy as relying on public acceptance comes to the fore. Sustainable development as transformation can itself be seen as a challenge to the legitimacy of the growth-oriented extractive economic system, proposing new criteria as legitimate.

The persistent historic inequalities in the distribution and experience of environmental harms (Ghosh, 2021) and the challenges to the legitimacy and effectiveness of governance for sustainable development stipulate the need for a multidimensional understanding of justice. Gupta et al. (2023) propose a framework for Earth System Justice that integrates intergenerational and multispecies justice, recognising the rights of non-humans as a part of the Earth system. Epistemic and recognition justice draw attention to the hegemony of the Western knowledge system, the limits that it poses for what can be considered valid knowledge, and the inclusion of a broader range of perspectives and potential human-nature interactions in governance (San Martín & Wood, 2021).

In governance for sustainable development as transformation, legitimacy can be bolstered through the application of a multidimensional understanding of justice and respect for the rights of humans and non-human species as well as the non-living environment. The rights-based approach in governance, the right to a clean environment, water or food presents one way of potentially evading problems, for example, generated by commercial, growth-oriented solutions and attendant new scarcities and inequalities (see above). Where the political will is present, this empowers the design of solutions that are not constrained by the need for profit making and accumulation, and that are genuinely aimed at regenerating the capacities and agencies of relevant human and non-human actors. As discussed in Chapter 6, the right to food aligns with food sovereignty and provides governments leverage to resist commercial interests in land acquisition for biofuels by obliging them to safeguard existing access to food and its means of production (Douwe van der ploeg, 2013, FIAN, 2020).

Both legitimacy and justice necessitate the integration of the lived experience as evidence informing governance design and interventions (Mcdermott & Surminski, 2018). Polycentric, networked governance constellations such as the Cities for Climate Protection (CCP) Programme, whereby constellations of formal and informal actors at various scales influence urban climate policy, have proven effective in such tricky situations where climate change mitigation seems to fly in the face of local livelihood opportunities (Bulkeley & Betsill, 2005). Deliberative, or strongly participatory governance, is generally seen as capable of affording innovative inclusion of plural forms of knowledge and supporting open discussion regarding conflicting interests and viewpoints. However, the outputs of deliberative governance and decision-making approaches are not always integrated into the policy process in a transparent or systematic manner (Boswell et al., 2023). Also, the way non-human entities should be integrated into governance remains less clear, but research is making inroads into this through approaches such as 'Umwelt cards', which represent the subjective world of an animal species and can be systematically consulted in decision-making (Maran, 2024).

7.2.2.4 Reflexivity in governance

Reflexive governance denotes the employment of broad, flexible and participatory problem-solving approaches that support learning (Voß & Kemp, 2006, Beck, 2006). It acknowledges that governance issues are situated in the context of complex systems and competing knowledge claims and therefore cannot be addressed through top-down predetermined solutions. Instead, what is needed is scope for the exploration of a broad set of alternatives and a diverse and flexible set of criteria for desirable outcomes. Therefore, reflexive governance encompasses activities aimed at learning, problem-solving and integration of plural knowledges. For example, it provides platforms for interaction that complement conventional governance decision-making, such as deliberative citizens' assemblies (Boswell et al., 2023) and foresight exercises such as participatory scenario planning (Waylen et al., 2015). It constantly evaluates and reconsiders societal ends – what the desired outcomes of governance are – against the background of diverse concepts and values. And it allows for experiments.

One example of such an experiment is the universal basic income-trial, often cited as a degrowth intervention (Liegey & Nelson, 2020), in Finland over two years from 2017 to 2019. Under the pilot project, 2000 unemployed Finns received €560 from the government. This unconditional payment was intended to enable them to establish meaningful livelihoods while relying on a guaranteed income from the state (BBC, 2019). In allowing for pilots and experiments, the approach to efficiency and effectiveness of reflexive governance differs from what is conventionally seen as good governance, namely the minimisation of time and cost of arriving at a predetermined solution to a societal problem (Beck, 2006). Reflexive governance involves experiments and deliberately integrates diverse viewpoints, even from actors who have limited capacities to articulate (the 'hard to reach'). Deliberation is not limted to policy formluation, but starts with the articulation of problem-definitions, to avoid non-decision-making (Chapter 4). Reflexive governance prioritises evidence gathering and decision-making practices where problem perceptions, assessment criteria and action strategies of different actors can be exposed to each other. While this may involve difficult conversations between or on behalf of those whose interests are in conflict or compromised in the course of addressing a societal problem, it enables learning - the acquisition of new knowledge, udnerstandings and values - on the part of all governance participants. This, in turn, can support change in both the broader understanding of how society works and in the values and knowledge that we use to define and understand our interests and to set personal goals.

7.3 Identifying and evaluating good governance

This section focuses on influencing change towards good governance through evaluation. Evaluation presents an important gateway for reflexivity and change. It forms an integral part of the governance cycle (Figure 7.1) and there is a vast literature on how to conduct governance or policy evaluation (Cairney, 2020). Ideally, regular evaluation exercises act as important evidence gathering and goal and process review exercises. In terms of the governance cycle depicted in Figure 7.1, evaluation usually takes place before the decision-making phase (ex-ante evaluation) and after the implementation phase (ex post evaluation) (Mickwitz, 2003a). Identifying and understanding good governance for sustainable development as transformation requires attention to governance practice, how governance (and conventional evaluation as a part of it) is conducted throughout the governance

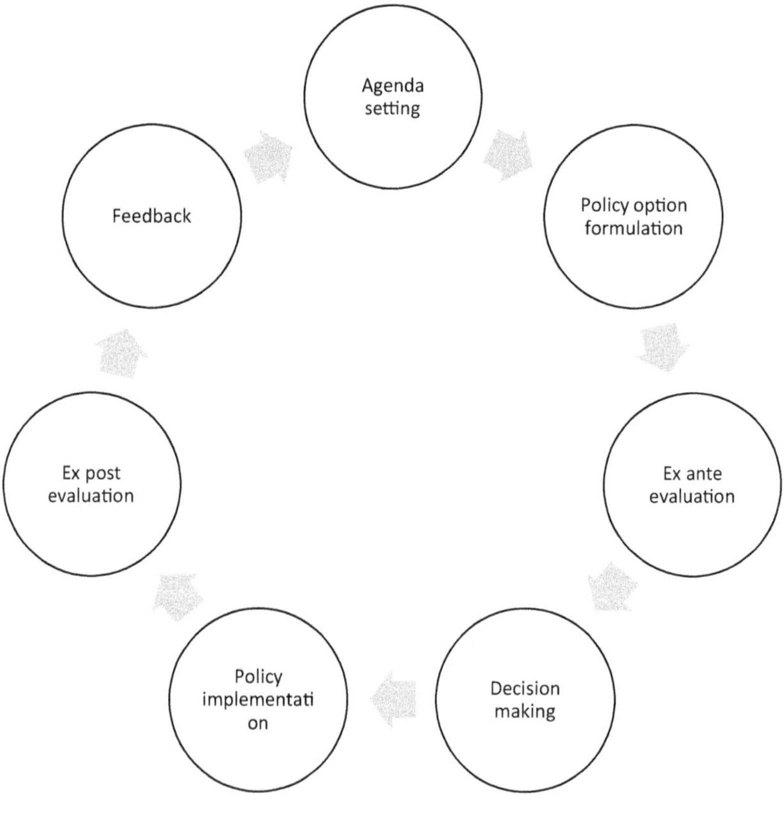

Figure 7.1 The governance cycle.

cycle, and governance outcomes and whether they align with sustainable transformation.

7.3.1 Evaluating governance practice

Integration, legitimacy and reflexivity are principles that pertain first and foremost to governance practice. Governance needs to integrate plural stakeholders and knowledges, embrace complexity, observe and manage inequalities in representation and participation, and adapt to changing circumstances and unexpected outcomes requiring frequent feedback and response loops beyond large evaluation exercises (Brousseau et al., 2012). This section focuses on the requirements the above principles set for how governance is done throughout the governance cycle.

We can think of governance 'starting' with agenda setting – an agreement on what issues merit a governance response in the first place and evidence gathering related to the extent and nature of the risks, drivers and stakeholders relevant to the issue. The agenda setting phase informs how the governance intervention is framed and designed (Morrison et al., 2019). At this stage already, the above principles are crucial. Legitimacy stems from procedural justice and the acceptability of governance goals to the public. The latter hinges on cultural norms and interests as well as formal justice – legislative requirements. Participation – whose voices are heard – and what kind of evidence is engaged with, are highly significant, and evoke procedural, recognition and epistemic justice considerations in the agenda setting phase (Juntti et al., 2009; San Martín & Wood, 2021). This is the phase where non-decision-making, narrow premises for evidence gathering or the lack of integration across relevant sectors would risk unexpected or biased outcomes and the legitimacy of the eventual governance intervention.

The framing of the governance issue to be addressed will inform the governance decision-making phase where decisions are made about intervention approaches and instruments (Morrison et al., 2019). For example, if the governance challenge is considered high risk, a regulatory approach may be adopted, where top-down targets and rules may be imposed. Alternatively, an economic instrument that does not prescribe action but respects local agency by offering incentives for certain behaviours may be selected (Evans & Thomas, 2024). Integration – coordination and alignment with other sectoral governance interventions is an important consideration here. Removing contradictions, realising synergies, and prioritising sustainability objectives across sectors should happen at this phase.

At this point of the governance cycle – when binding decisions are being made regarding what is deemed the best governance approach – ex-ante policy evaluation usually takes place (Figure 7.1). The term ex ante refers to 'evaluations of the possible impacts, in particular the environmental impacts, of new policies including the alternative of no action and the proposal for legislation and publication of the results' (Mickwitz, 2003a). Ex ante evaluation is an important evidence gathering and consultation exercise, where the principles of integration and reflexivity should be at the forefront. Participation is an important component of ex ante evaluation, as it informs the precise formulation of strategic or legislative goals and governance instruments and is therefore central to the legitimacy of governance itself.

Russel and Jordan (2008) suggest that rigorous ex ante evaluation should assist in the integration of environmental considerations into the core principles of all policies (not just those explicitly termed environmental policies) and therefore act as a key component of environmental policy integration across all sectors (Jordan, A. & Lenschow, 2010). While this is an important aspiration from the perspective of sustainable development, it is not easy to execute. Different policy sectors (from agriculture to industrial policy, for example) will have varying priorities and appraisal systems, which can create conflicts and inconsistencies (Russel et al., 2018). The lack of skilled personnel able to perform sustainability appraisals is also a common barrier. Effective communication strategies are needed to build awareness of sustainability concerns among communities and policy stakeholders and to foster their participation. Communication with other relevant policy sectors, perhaps across the food-water-energy nexus, is also important here.

Ex ante evaluation is integral to reflexive governance practice as it can help with the identification of unintended policy impacts at the decision-making stage, generate more and better quality information for intersectoral coordination and stakeholder collaboration and, if the participating formal and informal actors are open to it, support different types of learning that will lead to better, fairer and more effective governance design solutions. So-called 'instrumental learning' serves the specific intervention at hand and helps improve its workings in the specific environment in which it is being applied in (Russel & Jordan, 2008). Whereas 'conceptual learning' refers to the introduction of new ideas and perspectives into the policy process, most likely from deliberative consultation with stakeholders and experts relevant to the scope of the planned governance intervention (Russel & Jordan, 2008).

Conventional evaluation falls short of these considerations, placing significant emphasis on governance serving economic growth. In the UK, for example, all new state regulations need to be subject to an Options Assessment or an Impact Assessment, and in some cases both, before they can be passed by Parliament. An ex ante options assessment is required if the new regulation is forecast to have greater than +/-£10m annual net direct costs on business. Here, economic impact is therefore the primary consideration triggering a statutory ex ante evaluation. A final impact assessment will also be required if the impact on business is +/-£10m (Department for Business and Trade, 2023). The Better Regulation Framework does also feature the social impact of governance interventions, but the impacts of governance on sustainable development are to be considered only 'as appropriate and proportionate, for example distributional effects, net zero and environmental impacts' (Department for Business and Trade, 2023, p. 10).

Following the agenda setting and governance decision-making phases (Figure 7.1), governance can be seen to enter the implementation phase (Cairney, 2020). At this stage, it will also be decided how authority is to be delegated among the governance stakeholders – who is in charge of implementation and monitoring. This is the governance output (Chapter 4), where the designed new regulations and policies kick into action and the intended behavioural change and subsequent improvements should start to emerge. This is where polycentric governance constellations can be encouraged, enabling the integration of experiential knowledge and building further legitimacy in implementation.

Once the governance intervention has been in implementation for a period of time, the governance outcome, or the impact of governance, can be observed.

At this point then, ex post evaluation needs to take place. Whereas ex ante evaluation only happens once, ex post evaluation is something to be executed at regular intervals to ensure that governance is achieving the desired outcomes. Ex post evaluation refers to the evaluation of the effectiveness of existing measures in meeting set objectives (Mickwitz, 2003a). Ex post evaluation produces feedback for governance stakeholders, who should regularly return to the agenda-setting and decision-making phases to ensure that evaluation findings are integrated into the governance process and that governance keeps performing to the outlined principles. This sensitivity to feedback loops is central to reflexive governance, which is focused on modulating ongoing developments rather than controlling them, and thus needs to remain flexible, able to respond to emerging issues, such as newly diagnosed problems, inefficiencies, perspectives or knowledges (Voß & Kemp, 2006).

7.3.2 Evaluating governance goals, outputs, and outcomes

Mickwitz (2003a) suggests that ex post evaluation can take several different approaches. The 'goal-achievement model' (effectiveness evaluation) asks questions regarding whether the governance outcomes are in line with the goals established at the agenda-setting and decision-making phases of governance. Effectiveness evaluation overlooks potential side effects of the intervention in question, and the relevance of the goals themselves is not questioned. To address this oversight, the 'goal-free evaluation model' assesses the general outcome of an intervention, and the 'side-effects evaluation model' focuses specifically on the anticipated and unanticipated effects of evaluated interventions. When evaluating governance for sustainability transformation, evaluation should assess whether the outcomes really further a regenerative, planetary approach. But it remains important to also determine whether the outcomes are actually due to the governance intervention at hand. A systematic evaluation of a governance intervention should be informed by an understanding of the underlying theory of change (or intervention theory) describing how the intervention is intended to function.

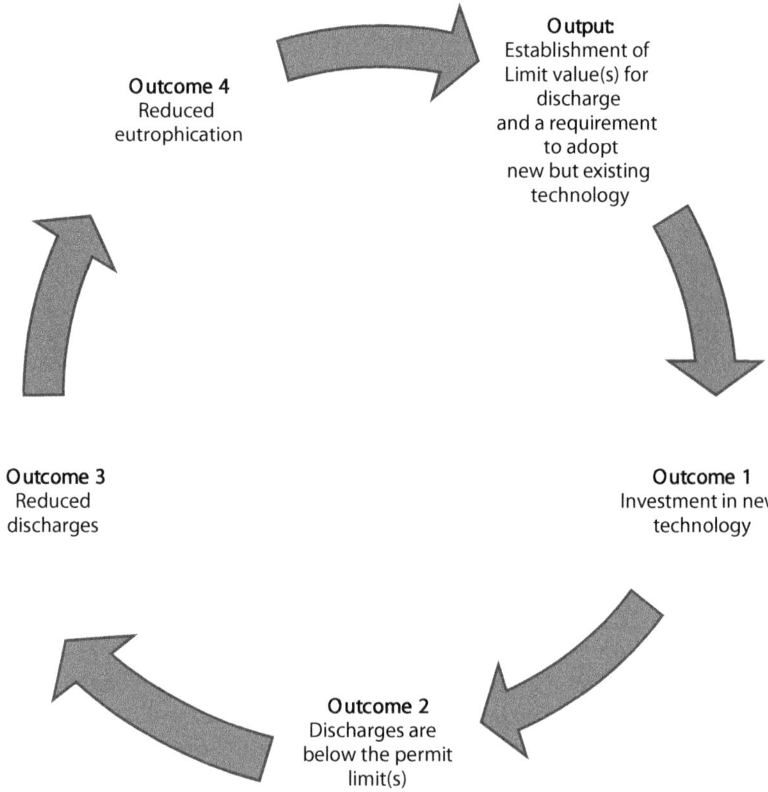

Figure 7.2 The intervention theory underpinning the key governance intervention aiming to mitigate the pollution emitted by Finnish pulp and paper mills. The intervention theory describes how the intervention, a permit-based system whereby each mill needs to first apply for and then regularly renew a permit for operation, is intended to work (Mickwitz, 2003b).

Intervention theories identify stakeholders, inputs, outputs, and outcomes of intended policies and point to the causal links which should be investigated in the evaluation (Figure 7.2).

Ex post evaluation should be conducted according to systematic evaluation criteria with a qualitative categorisation of effects (e.g. 'regenerative' vs 'detrimental' or 'environmental' and 'justice related') (Mickwitz, 2003a). An integrated approach should support the planetary perspective, directing focus across species and geographic locations, where effects are anticipated.

Mickwitz (2003a) suggests three categories of evaluation criteria for ex post evaluation of environmental policies including general, economic and social criteria. General criteria should focus on the 'relevance' of the intervention at hand. When evaluating governance aimed at transformational change this refers to whether the goals of the governance instrument address the key drivers of unsustainability underpinning the issue at hand and whether they are focused on achieving regenerative outcomes. For example, in terms of the ultimately unsuccessful Australian Clean Energy Mechanism above (Bailey, 2017, Chapter 5), the focus should be on whether the governance instrument – carbon pricing – addresses the key drivers of energy intensity. For example, in the food system, the key drivers of energy intensity pertain to the need to remain competitive in global markets, the scale of production and the level of mechanisation. In this context, setting a price on carbon emissions is unlikely to enjoy legitimacy or deliver on distributive justice.

Other general evaluation criteria proposed by Mickwitz (2003a) encompass effectiveness (to what degree do achieved outcomes correspond to intended goals of the policy) and persistence (regarding the longevity of governance outcomes). Flexibility (can the policy cope with changing conditions) relates directly to reflexivity – the ability of the governance intervention to adapt in response to feedback or changes in evidence and understanding regarding the issue it is trying to address. Predictability (the scope provided for those regulated and others to prepare and consider the policy and its implications) remains important for good governance in the era of sustainable development as transformation.

While cost-benefit (are the benefits worth the costs) and cost-effectiveness (do the results justify the resources used or could they be achieved with less expenditure) are widely used conventional evaluation criteria, they need to be re-shaped when evaluating governance outcomes in terms of sustainable development as transformation. This is because they reflect the prioritisation of economic considerations and the the contraints set by the growth imperative. Instead, when it comes to the implementation cost and economic impact of the governance intervention, these need to be weighed against the costs of a long-term no-action strategy. Questions need to be asked about whether the intervention encourages behaviour that is care oriented and regenerative, and helps to avoid economic losses resulting from the planetary crises if left unaddressed (Jackson, 2025).

Whereas the evaluation of governance practice directs attention to democracy-related criteria such as procedural and recognition justice (San Martín & Wood, 2021), evaluating the outputs and outcomes of governance directs attention to distributive justice. The planetary scope of governance for sustainable development as transformation stipulates a multispecies perspective. This can be realised, for example, by thinking in terms of the rights-based approach and what respecting the rights of waterbodies, animals, and plant species would mean in practice. This thinking should go beyond what could be perceived as an anthropocentric environmental conservation perspective focusing on species protection. We might need to ask questions about how the governance intervention impacts these different non-human actors' ability to function and to achieve and maintain wellbeing and integrity. For example, what are the opportunity costs for rivers when they are harnessed to serve the forest industry in terms of transport, waste assimilation, or power generation? Acknowledging their inherent sovereign existence is hoped to render rivers less vulnerable to exploitation rather than the predominant framing of rivers and other natural entities as forms of human property or public goods (Cohen et al., 2023)

Evaluation of governance outputs and outcomes also needs to probe legitimacy (to what degree is the policy instrument acceptable to stakeholders) and transparency (is the process, outcomes and instruments observable to outsiders) (Mickwitz, 2003a). A reflexive, sustainability focused approach stipulates a broader and more plural perspective on the goals and means of governance. It should ask, what is the transformational potential of this governance intervention, and does it advance just and legitimate transformation in the specified context.

7.4 Conclusions

The principles of good governance for transformational sustainable development need to be informed by the 21st century challenges and barriers to sustainability and transformative change. This chapter synthesises the key challenges as deriving one main cause: the economic growth imperative stipulated by the hegemonic global market economy. This in turn has led to the proliferation of complex systems that drive increasing reliance on technological solutions to achieve efficiency gains and the international configuration of commodity networks. The growth imperative also underpins a significant governance and implementation gap, meaning that sustainability targets, particularly pertaining to climate change, remain lacking in necessary ambition and fail in implementation. Finally, persistent inequality has led to a lack of legitimacy of both governance solutions and their knowledge basis.

Many so-called sustainable development solutions aim to maintain the extractive growth-oriented approach to planetary functions and resources, offering only incremental fixes to sustainability problems. Carbon taxes and a switch to biofuels without efforts to address the root causes of growing energy consumption represent examples of such partial fixes which fail to gain legitimacy and further exacerbate inequality. Such solutions do not address the growing inequalities and at worst, exacerbate them by producing new scarcities and favouring economies of scale. The prevalent populist surge threatening deliberative democratic decision-making structures and access to non-partisan information renders governance ever more prone to obfuscation by hostile stakeholders.

Sustainable development as societal transformation and emerging governance theorisation such as reflexive governance provide some steer for effective governance in this context. This chapter summarises this steer into four principles. The first one pertains to whether the governance intervention at hand is actually relevant to the underpinning drivers of unsustainability. The scope of the focus informing every governance intervention needs to be planetary. This stipulates some key messages such as the need for a regenerative approach that rewards activities that constitute care rather than extraction.

The appreciation of the Earth as a complex system, where the interdependencies between different domains of the environment and the economy need to be acknowledged and considered, also stipulates an integrated approach to governance – implying links across territorial boundaries and issues or sectors of governance. The food-water-energy nexus (Weitz et al., 2017) is an example of an integrated focus, whereas the Cities for Climate Protection (CCP) Programme exemplifies polycentric territorially networked governance constellations (Bulkeley & Betsill, 2005). The planetary focus also decentres humans as the key focus of governance and highlights our dependence on both non-human and non-living environments that have a right to exist and thrive, if not for any other reason than that we depend on them. In the context of a hugely unequal planet undergoing crises, all governance efforts need to be legitimate – that is, they need to serve just ends in a fair manner and need to be acceptable, perhaps as compromises, to the broad range of interests and perspectives that they implicate, from the human to the non-human, from Western to indigenous and from the present day to the future. Reflexivity, the final principle, really refers to the manner in which governance is undertaken. It stipulates a focus on learning by all stakeholders, formal and informal, throughout, reliance on plural knowledge, open discussion of conflicting values and interests, and flexibility, the ability to adapt following feedback from ongoing evaluation.

This chapter draws from examples in Chapters 5 and 6 to illustrate what the principles mean in practice. But there is no straightforward formula for getting governance right in the complex and varied global context, and an understanding of governance context and unexpected impacts is crucial for the realisation of the above four principles for governance that serves transformative sustainability. The final sections provide insights into how evaluation can support good governance for sustainable development as transformational change. In this endeavour, attention is required on both the 'hows' of governance practice and the 'whats' of governance outputs and outcomes.

References

Anderson, K. (2023, March 24). IPCC's conservative nature masks true scale of action needed to avert catastrophic climate change. *The Conversation*. Retrieved September 5, 2025, from http://theconversation.com/ipccs-conservative-nature-masks-true-scale-of-action-needed-to-avert-catastrophic-climate-change-202287

Bailey, I. (2017). Spatializing climate justice: Justice claim making and carbon pricing controversies in Australia. *Annals of the American Association of Geographers, 107*(5), 1128–1143. https://doi.org/10.1080/24694452.2017.1293497

Barrett, S. (2012). *The necessity of a multiscalar analysis of climate justice*. SAGE Publications. https://doi.org/10.1177/0309132512448270

BBC. (2019, 8th of February). Finland basic income trial left people 'happier but jobless'. https://www.bbc.co.uk/news/world-europe-47169549

Bulkeley, H. (2012). Governance and the geography of authority: Modalities of authorisation and the transnational governing of climate change. *Environment and Planning A, 44*(10), 2428–2444. https://doi.org/10.1068/a44678

Beck, U. (2006). Reflexive governance: Politics in the global risk society. In *Reflexive governance for sustainable development*. Edward Elgar Publishing. https://doi.org/10.4337/9781847200266.00011

Biermann, F., Van Driel, M., Vijge, M., & Peek, T. (2020). Governance fragmentation. In F. Biermann & R. Kim (Eds.), *Architectures of earth system governance: Institutional complexity and structural transformation*. Cambridge University Press.

Biermann, F., & Kim, R. (2020). Architectures of earth system governance: Setting the stage. In F. Biermann & R. Kim (Eds.), *Architectures of earth system governance: Institutional complexity and structural transformation* (pp. 1–34). Cambridge University Press.

Biermann, F., Hickmann, T., Sénit, CA. *et al.* (2022) Scientific evidence on the political impact of the Sustainable Development Goals. *Nature Sustainability 5*, 795–800. https://doi.org/10.1038/s41893-022-00909-5

Boswell, J., Dean, R., & Smith, G. (2023). Integrating citizen deliberation into climate governance: Lessons on robust design from six climate assemblies. *Public Administration*, *101*(1), 182. https://doi.org/10.1111/padm.12883

Brenner, N., & Elden, S. (2009). Henri Lefebvre on state, space, territory 1. *International Political Sociology*, *3*, 353–377.

Brousseau, E., Dedeurwaerdere, T., & Siebenhuner, B. (Eds.). (2012). *Introduction*. MIT Press.

Bulkeley, H., & Betsill, M. (2005). Rethinking sustainable cities: Multilevel governance and the 'urban' politics of climate change. *Environmental Politics*, *14*(1), 42–63. https://doi.org/10.1080/0964401042000310178

Cairney, P. (2020). *Understanding public policy: Theories and issues* (2nd ed.). Palgrave Macmillan.

Castells, M. (2004). *The network society: A cross-cultural perspective*. Edward Elgar Publishing Limited.

Chancel, L., Gómez-Carrera, R., Moshrif, R., & Piketty, T. (2026). *World inequality report 2026*. World Inequality Lab. https://wid.world/document/world-inequality-report-2026/

Clark, N., & Szerszynski, B. (2020). *Planetary social thought: The anthropocene challenge to the social sciences*. Polity Press.

Cohen, J. B., Dannreuther, C., Fraundorfer, M., Mackie, C., Martin-Ortega, J., Mdee, A., & Sutil, N. S. (2023). Riverkin: Seizing the moment to remake vital relations in the United Kingdom and beyond. *People and Nature, 5*(6), 1877–1892. https://doi.org/10.1002/pan3.10534

Coolsaet, B. (Ed.). (2020). *Environmental justice: Key issues*. Taylor & Francis Group.

Cumiskey, L., Priest, S., Klijn, F., & Juntti, M. (2019). A framework to assess integration in flood risk management: Implications for governance, policy and practice. *Ecology and Society*, *24*(4). https://doi.org/10.5751/ES-11298-240417

Darby, S. J., Wu, J., & Hammond, G. P. (2024). Briefing: Stocktaking global warming: The outcomes of the 2023 Dubai Climate Summit (COP28). *Proceedings of the Institution of Civil Engineers – Energy*. https://doi.org/10.1680/jener.24.00005

Department for Business and Trade. (2023). *Better regulation framework: Guidance*. UK Government, London. Retrieved May 29, 2026, from https://assets.publishing.service.gov.uk/media/67587ba55a2e4d4b993bfa83/better-regulation-framework-guidance-2023.pdf

Douwe Van Der Ploeg, J. (2013). Peasant-driven agricultural growth and food sovereignty. *The Journal of Peasant Studies*, *40*(6), 999–1030.

Evans, J. P., & Thomas, C. (2024). *Environmental governance* (2nd ed.). Routledge.

FIAN. (2020). *Right to adequate food and nutrition, and to food sovereignty*. FIAN. https://www.fian.org/en/right-to-adequate-food-and-nutrition-and-to-food-sovereignty-3/
Folke, C., Polasky, S., Rockström, J., Galaz, V., Westley, F., Lamont, M., Scheffer, M., Österblom, H., Carpenter, S. R., Chapin, F. S., Seto, K. C., Weber, E. U., Crona, B. I., Daily, G. C., Dasgupta, P., Gaffney, O., Gordon, L. J., Hoff, H., Levin, S. A., ... Walker, B. H. (2021). Our future in the Anthropocene biosphere. *Ambio, 50*(4), 834–869. https://doi.org/10.1007/s13280-021-01544-8
FAO, IFAD, UNICEF, WFP, & WHO. (2023). *The state of food security and nutrition in the world 2023. Urbanization, agrifood systems transformation and healthy diets across the rural–urban continuum*. FAO; IFAD; UNICEF; WFP; WHO. https://doi.org/10.4060/cc3017en
Ghosh, A. (2021). *The Nutmeg's Curse parables for a planet in crisis*. University of Chicago Press. https://doi.org/10.7208/chicago/9780226815466
Gupta, J., Liverman, D., Prodani, K., Aldunce, P., Bai, X., Broadgate, W., Ciobanu, D., Gifford, L., Gordon, C., Hurlbert, M., Inoue, C. Y. A., Jacobson, L., Kanie, N., Lade, S. J., Lenton, T. M., Obura, D., Okereke, C., Otto, I. M., Pereira, L., ... Verburg, P. H. (2023). Earth system justice needed to identify and live within Earth system boundaries. *Nature Sustainability*, *6*(6). https://doi.org/10.1038/s41893-023-01064-1
Infante-Amate, J., Travieso, E., & Aguilera, E. (2024). Unsustainable prosperity? Decoupling wellbeing, economic growth, and greenhouse gas emissions over the past 150 years. *World Development, 184*, 106754. https://doi.org/10.1016/j.worlddev.2024.106754
IPCC. (2023). *Climate change 2023: Synthesis report. Contribution of working groups I, II and III to the sixth assessment report of the intergovernmental panel on climate change* (Core Writing Team, H. Lee & J. Romero, Eds.). IPCC. https://www.ipcc.ch/report/ar6/syr/
Jackson, T. (2017). *Prosperity without growth: Foundations for the economy of tomorrow* (2nd ed.). Routledge. https://doi.org/10.4324/9781315677453
Jackson, T. (2025). *The care economy*. Polity.
Jordan. A., & Lenschow, A. (2010). Policy paper environmental policy integration: A state of the art review. *Environmental Policy and Governance*, *20*, 147–158. https://doi.org/10.1002/eet.539
Juntti, M., Özsezer-Kurnuç, S., & Dash, N. (2025). 'It's not necessarily a social space' – Institutions, power and nature's wellbeing benefits in the context of diverse inner-city neighbourhoods. *Landscape and Urban Planning*, *254*, 105241. https://doi.org/10.1016/j.landurbplan.2024.105241
Juntti, M., Russel, D., & Turnpenny, J. (2009). Evidence, politics and power in public policy for the environment. *Environmental Science & Policy*, *12*(3), 207–215. https://doi.org/10.1016/j.envsci.2008.12.007
King, A., & Cassidy, L. (2025). Earth is already shooting through the 1.5°C global warming limit, two major studies show. *The Conversation*. Retrieved September 30, 2025, from http://theconversation.com/earth-is-already-shooting-through-the-1-5-c-global-warming-limit-two-major-studies-show-249133
Koizumi, T. (2015). Biofuels and food security. *Renewable and Sustainable Energy Reviews*, *52*, 829–841. https://doi.org/10.1016/j.rser.2015.06.041
Kronenberg, J. (2025). From neoliberal urban green space production and consumption to urban greening as part of a degrowth agenda. *Cities*, *159*, 105739. https://doi.org/10.1016/j.cities.2025.105739
Kuipers, S. (2006). The crisis imperative. In *The crisis imperative* (pp. 9–18). Amsterdam University Press. https://doi.org/10.1515/9789048503926-002
Lélé, S. M. (1991). Sustainable development: A critical review. *World Development*, *19*(6), 607–621. https://doi.org/10.1016/0305-750X(91)90197-P

Liegey, V., & Nelson, A. (2020). *Exploring degrowth: A critical guide* (1st ed.). Pluto Press.
Maran, T. (2024). *Using Umwelt cards to give non-humans a voice*. Retrieved July 3, 2025, from https://co-evolvers.eu/using-umwelt-cards-to-give-non-humans-a-voice
Marsden, T., Faus, A., & Sonnino, R. (2019). Reproducing vulnerabilities in agri-food systems: Tracing the links between governance, financialization, and vulnerability in Europe post 2007–2008. *Journal of Agrarian Change, 19*(1), 82–100. https://doi.org/10.1111/joac.12267
Mcdermott, T. K. J., & Surminski, S. (2018). *How normative interpretations of climate risk assessment affect local decision-making: An exploratory study at the city scale in Cork, Ireland*. The Royal Society. https://doi.org/10.1098/rsta.2017.0300
McDonagh, J. (2014). Rural geography II: Discourses of food and sustainable rural futures. *Progress in Human Geography, 38*(6), 838–844. https://doi.org/10.1177/0309132513514507
McGreevy, S. R., Rupprecht, C. D. D., Niles, D., Wiek, A., Carolan, M., Kallis, G., Kantamaturapoj, K., Mangnus, A., Jehlička, P., Taherzadeh, O., Sahakian, M., Chabay, I., Colby, A., Vivero-Pol, J., Chaudhuri, R., Spiegelberg, M., Kobayashi, M., Balázs, B., Tsuchiya, K., ... Tachikawa, M. (2022). Sustainable agrifood systems for a post-growth world. *Nature Sustainability, 5*(12), 1011–1017. https://doi.org/10.1038/s41893-022-00933-5
Mickwitz, P. (2003a). A framework for evaluating environmental policy instruments: Context and key concepts. *Evaluation*, *9*(4), 415–436.
Mickwitz, P. (2003b). Is it as bad as it sounds or as good as it looks? Experiences of Finnish water discharge limits. *Ecological Economics*, *45*, 237–254. https://doi.org/10.1016/S0921-8009(03)00081-8
Morrison, T. H., Adger, W. N., Brown, K., Lemos, M. C., Huitema, D., Phelps, J., Evans, L., Cohen, P., Song, A. M., Turner, R., Quinn, T., & Hughes, T. P. (2019). The black box of power in polycentric environmental governance. *Global Environmental Change*, *57*, 101934. https://doi.org/10.1016/j.gloenvcha.2019.101934
O'Neill, J. (2017). Life beyond capital | Essay by John O'Neill. https://cusp.ac.uk/themes/nature/m1-6/
Oxfam. (2023). *Climate equality: A planet for the 99%*. Oxfam GB. https://doi.org/10.21201/2023.000001
Piketty, T. (2014). *Capital in the twenty-first century*. Harvard University Press.
Raworth, K. (2017). *Doughnut economics: Seven ways to think like a 21st century economist* (1st ed.). Random House Business Books.
Ray Biswas, R., & Rahman, A. (2023). Adaptation to climate change: A study on regional climate change adaptation policy and practice framework. *Journal of Environmental Management, 336*, 117666. https://doi.org/10.1016/j.jenvman.2023.117666
Richardson, R., Steffen, S., Lucht, L., Bendtsen, B., Cornell, C., Donges, D., Drüke Drüke, Fetzer, F., Bala, B., von Bloh, v. B., Feulner, F., Fiedler, F., Gerten, G., Gleeson, G., Hofmann, H., Huiskamp, H., Kummu, K., Mohan, M., Nogués-Bravo, D., ... Rockström, J. (2023). Earth beyond six of nine planetary boundaries. *Science Advances*, *9*(37), eadh2458. https://doi.org/10.1126/sciadv.adh2458
Russel, D., Turnpenny, J., & Jordan, A. (2018). Mainstreaming the environment through appraisal: Integrative governance or logics of disintegration? *Environment and Planning C, 36*(8), 1355–1370. https://doi.org/10.1177/2399654418767656
Russel, D., & Jordan, A. (2008). Gearing-up governance for sustainable development: Patterns of policy appraisal in UK central government. *Journal of Environmental Planning and Management*, *50*(1), 1. https://doi.org/10.1080/09640560601048267

San Martín, W., & Wood, N. (2021). Pluralising planetary justice beyond the North-South divide: Recentring procedural, epistemic, and recognition-based justice in earth-systems governance. *Environmental Science & Policy*, *128*, 256. https://doi.org/10.1016/j.envsci.2021.12.002

Steffen, W., Richardson, K., Rockström, J., Cornell, S. E., Fetzer, I., Bennett, E. M., Biggs, R., Carpenter, S. R., de Vries, W., de Wit, C. A., Folke, C., Gerten, D., Heinke, J., Mace, G. M., Persson, L. M., Ramanathan, V., Reyers, B., & Sörlin, S. (2015). Planetary boundaries: Guiding human development on a changing planet. *Science, 347*(6223), 1259855. https://doi.org/10.1126/science.1259855

Sze, J. (2020). *Environmental justice in a monent of danger*. University of California Press.

Sørensen, E. (2002). Democratic theory and network governance. *Administrative Theory & Praxis*, *24*(4), 693–720. https://doi.org/10.1080/10841806.2002.11029383

UNFCCC. (2024). *The Paris Agreement*. Retrieved March 12, 2024, from https://unfccc.int/process-and-meetings/the-paris-agreement

Voß, J. P., & Kemp, R. (2006). Sustainability and reflexive government: Introduction. In *Reflexive governance for sustainable development*. Edward Elgar Publishing. https://doi.org/10.4337/9781847200266.00009

Voulvoulis, N., Giakoumis, T., Hunt, C., Kioupi, V., Petrou, N., Souliotis, I., Vaghela, C., & Wan Rosely, W. (2022). Systems thinking as a paradigm shift for sustainability transformation. *Global Environmental Change, 75*, 102544. https://doi.org/10.1016/j.gloenvcha.2022.102544

Waylen, K. A., Martin-Ortega, J., Blackstock, K. L., Brown, I., Avendaño Uribe, B. E., Basurto Hernández, S., Bertoni, M. B., Bustos, M. L., Cruz Bayer, A. X., Escalante Semerena, R. I., Farah Quijano, M. A., Ferrelli, F., Fidalgo, G. L., Hernández López, I., Huamantinco Cisneros, M. A., London, S., Maya Vélez, D. L., Ocampo-Díaz, P. N., Ortiz Guerrero, C. E., Pascale, J. C., ... Zilio, M. (2015). Can scenario-planning support community-based natural resource management? Experiences from three countries in Latin America. *Ecology and Society, 20*(4), Article 28. https://doi.org/10.5751/ES-07926-200428

WEF. (2025). *The global risks report 2025* (20th ed.). World Economic Forum. Global Risks Report 2025 PDF

Weiss, T. G., & Wilkinson, R. (2018). *From international organization to global governance* (T. Weiss & R. Wilkinson, Eds., 2nd ed.). Routledge, Taylor & Francis Group.

Weitz, N., Strambo, C., Kemp-Benedict, E., & Nilsson, M. (2017). Closing the governance gaps in the water-energy-food nexus: Insights from integrative governance. *Global Environmental Change*, *45*, 165–173. https://doi.org/10.1016/j.gloenvcha.2017.06.006

Whatmore, S. (2006). Materialist returns: Practising cultural geography in and for a more-than-human world. *Cultural Geographies*, *13*(4), 600. https://doi.org/10.1191/1474474006cgj377oa

Wilson, G. A., & Burton, R. J. F. (2015). 'Neo-productivist' agriculture: Spatio-temporal versus structuralist perspectives. *Journal of Rural Studies, 38*, 52–64. https://doi.org/10.1016/j.jrurstud.2015.02.003

World Bank Group. (2025). *June 2025 update to global poverty lines*. World Bank Group. Retrieved December 4, 2025, from https://www.worldbank.org/en/news/factsheet/2025/06/05/june-2025-update-to-global-poverty-lines

8 Conclusions

Learning objectives

- What do the principles for good governance for transformative sustainable development mean for the future direction of governance?
- How can climate assemblies address the legitimacy deficit and the implementation gap?
- How can alternative food system governance approaches support a regenerative approach and justice in the food system?
- What scope is there for governance to influence change within the dominant economic paradigm of the global capitalist market economy?

8.1 Introduction

The main aim of this book is to explore the key challenges and principles for good governance for sustainable development in the 21st-century multi-crises context. Coined in the 1980s, sustainable development has enjoyed a revival on the global agenda since the 2015 launch of UN's Agenda 2030 and the Sustainable Development Goals (SGDs) (UN General Assembly, 2015). But despite its 40-year lifespan and numerous global conventions such as the UN Environment and Development Programme, Agenda 21 and, at the time of writing, Agenda 2030, not much progress has been achieved in the name of sustainable development (Biermann et al., 2022; Chancel et al., 2026; Oxfam, 2023; Redclift, 2005). Presently, the world is headed for a climate breakdown which, in the context of persistent and growing inequality and lack of appropriate means to address its complex and uneven impacts, is set to have a potentially catastrophic impact on humankind (Anderson, 2023; IPCC, 2023).

As discussed in Chapter 7, the slow progress of sustainable development is largely due to the dominant normative agenda that prioritises economic growth and a subsequent lack of political will, the complexity of the required system transformation, and the reliance on incremental sustainability solutions which, in the current context, often risk further disadvantaging the poor. Governance solutions that aspire to operate in the context of the current economic system pose difficult trade-offs, appearing either as costs to

DOI: 10.4324/9781003179207-11

businesses and households or as exclusive commodified solutions that drive inequality. For example, integrating GHG emissions into the economy as costs risks compromising the viability of businesses and jobs. Moreover, unless supported by significant government subsidies and regulation, renewable energy technologies such as heat source pumps and electric cars remain out of reach of poorest households and, at worst, due to policy fragmentation and issue leakage, cause environmental and social harms in the supply chain.

This book stages its exploration of good governance for transformational sustainable development in the context of the prevalent multi-crises of planetary instability, socio-economic inequality and highly politicised knowledge. In Part 1 'Setting the scene – governing in the age of crises', Chapters 1 and 2 explored how the crises context renders governance a challenging endeavour. Chapter 3 discussed sustainable development itself, identifying two potentially complementary approaches. The green economy aims to mitigate environmental harms in the context of the dominant economic paradigm while degrowth stipulates a radical cultural shift away from its key values and principles. The definition of sustainable development for the purpose of this book was set out as *fundamental societal transformation towards equitable and regenerative relations among people and the planet, addressing the core drivers of unsustainability*. Chapter 4 explored useful governance theorisations, pointing to the notion of relational power and networked, reflexive governance arrangements as supportive of a transformational shift towards sustainability. In Part 2 'Understanding global governance issues and solutions', Chapters 5 and 6 explored climate change and food security governance approaches and examples. Using the theories explored in Chapters 3 and 4, Part 2 examined the challenges that current policies are facing as well as opportunities for progress towards transformational sustainability in those two issue areas. The success of climate policy was seen to hinge on the perceived legitimacy of interventions such as carbon pricing, requiring sensitivity to how climate justice is understood at various scales of governance. Policy integration across sectors and countries emerged as crucial. Food sovereignty - the agency that farmers, communities and national governments hold over how food is produced, priced and distributed - was established as key to food security but difficult to uphold in the global food system. In this context, framing food in terms of a right rather than simply a commodity, could provide governments the leverage to defend local food security in the face of corporate interests. Finally, in Part 3 'Lessons for good governance for sustainable development', Chapter 7 provided a synthesis of the barriers to societal transformation for sustainability and set out a range of principles to guide governance in navigating and addressing them. The chapter also applied the principles to the evaluation of the practice and outcomes of governance.

8.2 Transformational governance for climate mitigation and adaptation

As seen in Chapter 5, climate change makes for a wicked governance issue that calls for the integration of both scientific and experiential knowledge - an understanding of the way that climate change drivers, impacts and governance interventions are experienced in practice. In order to achieve legitimacy of both governance instruments and their outcomes, governance needs to take climate justice seriously. The fate of carbon pricing in Australia discussed in previous chapters underlines this. The yellow vest protests in France in 2018 further exemplify public resistance to unjust governance interventions. The country-wide protests were triggered by a GHG tax on petrol that unfairly disadvantaged the already hard up, rather than the rejection of climate mitigation as a governance goal (Driscoll, 2023). Legitimacy in climate governance hinges on how responsibility for causes and mitigation measures, and the ability to adapt to climate change are perceived and how the resulting regulatory burden is distributed in relation to this.

The international regimes, multilevel governance and Earth system governance theories are useful lenses through which to critically view climate governance at the global scale. The constructivist lens helps to highlight how global climate justice is reflected in the values and norms of global agreements. Presently, for example, the Nationally Determined Contributions to climate change (NDCs) and Net Inventory Reports (NIRs) do not reflect the historical legacy of each country in causing climate change. The US is responsible for the biggest historical emissions, more than 1.5 times those of China, in second place, closely followed by the UK and the 27 EU countries (Ritchie & Roser, 2024). Neither do territorial emissions reported in the NIRs reflect the high emissions caused by consumption inherent to Western lifestyles. While developed nations like the UK and the EU of 27 have made significant reductions in annual territorial CO_{2e} emissions in the recent past, the high-consumption lifestyles led by their populations mean that they persist as the most significant drivers of climate change today and account for much of China's emissions. These hidden responsibilities underpin the calls for restitution justice especially by small island states whose territorial integrity is under threat form climate change, and undermine the realisation of climate justice in global climate governance.

That said, climate justice is framed and experienced differentially across governance scales. National and local scale interventions implementing global treaties implicate a number of economic activities and sectors and planning and sectoral policies at national and sub-national scales. It needs to advance in an integrated manner and build legitimacy through community and stakeholder engagement informed by the everyday experience of climate change related risks and responsibilities. Governance for emission mitigation needs to be alive to both global and local scale climate justice, integrating considerations for distributive, procedural and recognition justice (Coolsaet,

2020). Adaptation policy in turn needs to address the underlying causes of climate vulnerability vested in societal structures that uphold inequalities and the ability of communities, organisations and individuals to cope with climate change impacts. Reflexive governance approaches are crucial for building legitimacy at the local scale so that international commitments can be successfully implemented into practice. Citizens' assemblies provide a model that, if well designed, they afford opportunities for integrated and reflexive governance in practice.

8.2.1 Citizens' assembly for climate governance in France

The crisis of knowledge and rise of populism (Chapter 2) manifest in climate scepticism and outright rejection of the need for climate mitigation among some political leaders as well as communities and individuals. The French yellow vest protest spoke to the difficulty of addressing climate justice in the context of the current economic system and associated inequality and precarity (Driscoll, 2023). But the legitimacy of climate policies has been further undermined by the emergence of populism – manipulation of information with the aim to simply acquire political power – and an associated corrosion of democratic structures and values in the 21st century (Grant, 2021).

This polarisation of socio-economic fortunes and politicisation of knoweldge highlights the crucial role of justice and legitimacy in governing for sustainable transformation. In the context of sustainable development, considerations of distributive and procedural justice need to be accompanied by notions of recognition and reparation justice, all of which extend beyond the human perspective to encompass other species and the planet's functional capacities (Gupta et al., 2023). This requires reflexivity – the ability of governance to integrate, contrast and, where possible, reconcile differing perspectives, values and knowledges. One way of doing this is through citizens' assemblies (Boswell et al., 2022). Boswell et al. (2022) describe citizens' assemblies as deliberative 'mini publics' aiming to integrate citizens' experiential knowledge into governance alongside expert and technocratic perspectives. Ideally, they enable conflict resolution by bringing together diverse publics and affording unconstrained and authentic deliberation, which has the potential to support learning, and changes in perceived interests and values as well as nurture innovative capacity (Bulkeley & Mol, 2003).

The potential for deliberative governance to achieve transformational change has convinced, for example, the direct-action group Extinction Rebellion (XR) (Grossman, 2020) to shift focus towards a more collaborative approach to achieving change. XR is campaigning for citizens' assemblies on climate and ecological justice, where members are set the task of collaborating to investigate and find solutions. To support this, they hear balanced information from experts, stakeholders and people already affected by the crisis (Extinction Rebellion, 2025). In citizens' assemblies, a random selection of people is invited to receive information, ask questions, discuss, and debate regarding governance interventions relating to climate mitigation and adaptation. They thus provide good potential for both integration across sectoral and issue boundaries and reflexivity – the incorporation of plural values and knowledge and the contrasting of conflicting views and interests (Beck, 2006).

Cherry et al. (2021) describe the process and outcomes of the 'Convention Citoyenne pour le Climat' (CCC), a citizens' assembly mandated directly by President Emmanuel Macron in France between 2019 and 2020. The 150 randomly selected members were asked how France could reduce greenhouse gas (GHG) emissions by at least 40% by 2030 in a spirit of social justice. The mandate explicitly promised that proposals would be forwarded 'without filter' to referendum, parliamentary vote, or regulation—creating strong political commitment and expectations on the part of the participants. Participants were ordinary citizens, not self-selected volunteers. The assembly met seven times in in-person sessions for a 2.5-day duration each time. Several online sessions and optional webinars were also offered to ensure that participants had access to expert knowledge and an understanding of governance. The CCC followed a strongly bottom-up method, whereby citizens generated policy ideas themselves, rather than choosing from options pre-defined by experts. The measures proposed by the assembly evolved through brainstorming, prioritisation, redrafting, and negotiation over several sessions. Experts were present, but they supported rather than directed deliberation by providing technical input and feedback on feasibility and helping translate proposals into legally viable measures. The emphasis was on citizens' self-organisation and they ended up working in five groups of 30 focusing on different aspects of the necessary governance response to GHG emission reductions.

The output of this deliberative reflexive governance activity was 149 concrete measures, grouped into thematic blocks. Cherry et al. (2021, p. 55) focus on the outputs from the group that worked on the theme of consumption. The outcome of deliberations of that group was five measures intended for the Climate and Resilience Bill:

- Create an obligation to disclose the carbon impact of products and services.
- Regulate advertising to reduce incentives for overconsumption.
- Limit overpackaging and the use of single-use plastics by developing bulk products and deposit schemes in distribution sites.
- Make education, training and awareness-raising the leverage tools for responsible consumption.
- Ensure better implementation of public environmental policies and evaluate them in order to make them more effective.

The presidential remit guaranteed high-level political commitment, but Cherry et al. (2021) suggest that this may have foregrounded unrealistically high expectations among the participants. Although many proposals entered the Climate and Resilience Bill, the participating citizens ultimately judged government uptake as partial and disappointing, illustrating tensions between deliberation and political translation into lasting governance outcomes. This reflects the findings of Boswell et al. (2022) which point to limitations in the ability of formal governance structures to integrate deliberative and reflexive governance practices such as citizens' assemblies. The lack of political will can therefore continue to hinder reflexive governance efforts even if they have gained strong legitimacy. Nevertheless, a significant outcome of the CCC was the measurable increase in public awareness of climate policy (Cherry et al., 2021). The assembly also led to the establishment of 'Les 150', a civil-society association to monitor government implementation of the proposed measures. These deliberative forums serve an important purpose in overcoming climate scepticism among communities and hold potential for innovative solutions that also further the cultural change necessary for sustainability as transformation (Chapter 3).

8.3 Food system transformation

In Chapters 3 and 7 it was suggested that good governance for sustainable development hinges on a regenerative (rather than extractive) approach to the planet, respecting the rights of humans, other species and natural entities such as rivers and forests. In the context of the international food system, this refers to the environmental sustainability and climate resilience of food production, the rights of farmers and agricultural workers in terms of access to viable livelihoods, and the ability of people and communities to make informed and sustainable choices when it comes to the food that they consume. The notion of relational power points to some useful guidance for governing such wicked issues (Chapter 4). For example, it highlights the importance of networked, plural governance constellations and reflexive, critical interrogation of prevalent institutional norms and inclusion of different types of evidence.

The relational understanding of power integrates non-human and inanimate entities such as technological solutions into the governance network as imbued with agency. Relational power enables us to understand the tacit agency of non-human entities, such as other species, ecosystems and natural resources on which we depend. This non-human rights perspective supports the regenerative care-oriented approach to the planetary system; if we accept that the planetary system encompasses non-human actors who also have rights, then governance needs to serve their interests as well as those of humans. Cohen et al. (2023) explore this shift in perspective through the notion of kinship with natural entities such as rivers, for example. Granting rivers agency and rights resembles the way Indigenous knowledge integrates nature as 'kin' – a relation that we should reciprocally nurture, which should then inform legislation and water governance (Cohen et al., 2023).

System complexity renders conventional, solution-oriented approaches in both climate change and food security governance difficult. Presently, the global industrial food system prioritises competitive large-scale production by corporate land owners and places agency in the hands of large multinational supermarkets and agri-tech companies in setting both farm-gate and consumer prices. These corporate stakehodlers are often able to evade regulation by national governments and their growth oriented competitve rationale underpins a constant search for comparative advantage and economies of scale, which means that food often travels long distances before it reaches our plates. This has led to the weakening of consumer agency in terms of scope for ethical decision-making and the marginalisation of small-scale producers who constitute the majority of farmers worldwide (Ritchie, 2021). While sustainable food production practices such as agro-ecological practice or organic farming have good potnetial to nurture a rengerative approach to soil and biodiversity, they do not align with the competitive rational of productivist agriculture and are currently being advanced through incremental governance initiatives such as subsidies and product labelling. More transformational changes are driven mainly by informal governance actors that aspire to secure small-scale access to land and reinstate the connection between consumers and producers for increased transparency and accountability in the food system. One example of this is the Monmouthshire Food Partnership in Wales, UK.

8.3.1 A local food network advancing transformational sustainability in the UK

The prevalent socio-technical systems, such as the international food system, are energy intensive and foster an extractive approach to resources and labour. In the case of the food system, this is associated with large scale intensive food production causing profound alterations to land cover and the land system itself with severe implications for ecosystem functions, habitats and species (Steffen et al., 2015). The extent of resource appropriation and transformation into energy, consumer goods and food and its by-products such as solid waste and emissions into water and the atmosphere are now impacting every aspect of the planet (Folke et al., 2021). Moreover, the global food system mirrors colonial power relations, where small farms especially in developing countries struggle for market access and land rights. This has compromised local food security in many disadvantaged contexts where the lack of support structures has left small farms unable to cope with challenges posed by climate change or conflict. Meanwhile, consumers in more affluent regions depend on long and complex supply chains and large corporate retailers and have little knowledge or agency over the origin and production methods of foods that they have access to. Ethically produced and sustainable products are sold at a premium and only accessible to well-off middle-class consumers. These developments have begun to cause vulnerability in the food system itself (Marsden et al., 2019).

But there are initiatives aiming to counter this corporatisation and globalisation of food production and supply. Scott et al. (2025) explore how Our Food 1200, a community benefit society, and the local Monmouthshire County Council contribute to the revival of a small-scale local food system while instilling a regenerative and just ethos to food production and access in south Wales, UK (Our Food, 2025; Sustainable Food Places, 2025). They represent a networked partnership model of governance which engages a broad scope of local stakeholders such as landowners, farmers, manufacturers and retailers, supporting both legitimacy and fairness in the governance process (Scott et al., 2025). This governance configuration originated from a five-year project (Sustainable Food Places 2019–2024) led by three non-governmental organisations, demonstrating that sometimes change can start with small and temporary solutions. The project kick-started the Monmouthshire Food Partnership, which currently lists its aims as: increasing the sustainability of food supply and access to healthy local food; structural transformation of the local agri-food economy in favour of small-scale locally embedded production; increased access to land by those with skills and willingness to produce sustainably with attention to multifunctionality and environmental assets; better local business viability and diversity; and increased access to the knowledge and skills necessary for healthy and sustainable diets and production methods.

Our Food 1200 constitutes an alternative model of land governance enabling a small-scale production structure that attracts producers whose key aim is to produce sustainably through agro-ecological and related practices, for the local market. This governance model has the potential to disrupt the ongoing commodification of land and the surging land prices barring access to land for small farmers and driving inequality in the food system. The main vehicle of Our Food 1200 is the Farmland Trust, a not-for-profit community land trust, which acquires farmland and converts it into small farms and dwellings for experienced farmers interested in establishing sustainable production systems that serve local needs. This is an innovative solution countering the push for consolidation of production resulting from the cost-price squeeze and the financialisation of land in the global food system, that favours large-scale corporate landownership (Jackson et al., 2021).

The Monmouthshire Food Partnership (Sustainable Food Places, 2025), in turn, addresses sustainability on the manufacturing and demand side of the food system. It consolidates the County Council's legacy on mitigating food poverty and integrates stakeholders from industry as well as communities into addressing the access and utilisation pillars of food security to combat inequality and injustice in the food system (Jones & Hill, 2023). The partnership advocates for local and national scale policy to support a more sustainable, small-scale food system and raises awareness and engagement around healthy and sustainable food within communities. It aims to also tackle food poverty while offering a networking platform, training, and events to support sustainable local food businesses and sustainable food procurement.

This governance network consists mainly of informal governance actors such as the land trust, local food producers and businesses, and community groups. But formal institutions such as the Future Wellbeing of Future Generations Wales 2015 Act and the Healthy Weight, Healthy Wales obesity strategy provide formal legitimacy, allowing these bottom-up initiatives to thrive (Jones & Hill, 2023). Oversight, knowledge and network brokering from Food Sense Wales (a charitable health sector organisation) and a recent Welsh government award for local authorities to support the development of local food partnerships have played an important role also. Powys County Council has ensured that Local Development Plans provide an opportunity to integrate such measures into planning more broadly (Scott et al., 2024). The work of the Monmouthshire Food Partnership and Our Food 1200 serves as an example of deriving wellbeing, liveability and food security from a balanced stock of rural assets without an economic growth imperative (Jackson, 2017). This governance network therefore shows potential to initiate transformative change in the global food system, with attendant benefits to both community and land-based assets and the way they are engaged with to produce local food security in Southeast Wales.

8.4 Governing for sustainable development in the growth-driven global market economy

The above two examples of governance for sustainable development as transformation stop short of challengeing the hegemony of the global capitalist market economy or indeed even directly questioning the need for economic growth. However, they do both further and represent a transformation of power structures, priorities and practices associated with it and challenge the prioritisation of the growth imperative in various ways. For example, in safeguarding access to farmland through the non-profit Farmland Trust, Our Food 1200 challenges the financialisation of land (Marsden et al., 2019), which sees land prices steadily rising, further contributing to the cost-price squeeze at farm level, and bolsters access to land by providing an institutionalised model for small-scale land ownership as a value in itself (Gallent et al., 2015). Similarly, in demanding the regulation of advertising, education for responsible consumption and more effective implementation of environmental policies in France, the CCC challenges the notion of consumption-driven economic growth at any cost and promotes a change in societal values.

Both governance initiatives are therefore promoting the kind of cultural change that is central to degrowth and a more regenerative relationship with the planet (Jackson, 2025). Both work to increase agency at the local scale. The citizens' assemblies directly promote equity by advancing each person's right to be heard through non-hierarchical modes of organisation (Liegey & Nelson, 2020). They offer new institutions and structures that target dependencies that currently tie us into growth (Jackson, 2025). In promoting a regenerative approach to land in farming and connectivity between like-minded local food businesses, the Monmouthshire Food Partnership enables collective behaviours aimed at increasing access to food, collective food production and processing and avoiding food waste, all of which are activities that align with a degrowth ideology of 'frugal abundance' (Liegey & Nelson, 2020). An important function of the food partnership is reconnecting consumers with food producers and manufacturers and re-embedding food in the local culture and landscape (Douwe Van Der Ploeg, 2013). In so doing, it also implements many of the principles of the right to food (Chapter 6).

Contrary to common belief, most interpretations and manifestations of degrowth do not demand a 'shrinking' of the economy but focus on decentring the growth imperative. Liegey and Nelson (2020) emphasise resistance to extractive processes, institutions and imaginaries and a carving of space for alternatives which escape subjugation to the growth imperative via engaging alternative forms of collaboration, exchange and care. Initally, committing more time towards care may not seem an attractive idea signalling better quality of life and wellbeing from everyone's perspective. Perhaps this is because care has been framed in such an unfavourable light in contemporary society. Care workers were celebrated during the Covid-19 pandemic but many essential care providers remain underpaid in comparison to other professions. Care work is seen as hard and requiring self-sacrifice. But care for human health and, as science attests, for the planet is after all core to the functioning of social systems, cohesive communities, peaceful societies and viable economies. Jackson (2025) therefore suggests that a revalorisation of care is needed, as what economies 'see' as productive work is after all a value-based decision. The provision of care enabling human systems as well as the planet to maintain a favourable equilibrium should therefore be seen and valued as an investment, rather than a cost.

Perhaps a revalorisation of care can already be seen in the rise in attention to wellbeing, care for oneself, with a Western wellbeing industry drawing on a range of non-Western knowledge systems and ideologies such as Ayurvedic science from India or rituals and practices of connectivity and integration with nature from different Indigenous cultures. In his book, *The Care Economy* (2025, p. 229), Tim Jackson defines care as 'activities to maintain, restore and repair health', that should extend from the individual scale to communities and the planet. He suggests that currently, the capitalist market economy is blind to the quality of economic processes whereby its engine, profit, is extracted from humans, technology and the planet's many functions and resources. As a result, the economic system casts activities that in fact diminish wellbeing of both people and the planet as desirable and even necessary. What needs to happen is a qualitative turn where harmful activities do not count as growth. He uses as an example the way in which profit is derived from the manufacturing and sale of ultra-processed foods that carry a high risk of causing obesity and chronic diseases such as diabetes. Further profit is then extracted from the pharmaceutical industry from the manufacturing and sale of medicines that treat these diseases essentially stemming from over-consumption. This is all portrayed as positive growth in terms of GDP and corporate turnover. While governments currently experiment with disincentives such as sugar taxes and bans on advertising 'junk foods', even such incremental 'breaks' on the growth machine face stark opposition from those who benefit from the associated profits. A fundamental reframing of the value of economic activities is needed to incentivise a shift from extraction to care.

Some forward-looking economic thinkers such as Jussi Lindgren (2024) suggest that the capitalist market economy may already be in evolution towards a more care-oriented version. Climate breakdown and the rise of national debt are coinciding with unplanned and largely unmanaged economic stagnation across a range of developed economies. This situation alone is difficult to manage without compromising the spending power of a large proportion of low- to middle-income households, as seen in the cost-of-living crisis of the early 2020s. As technology, including Artificial Intelligence, provides increasing efficiency gains in industry and service provision, reducing employment opportunities even in highly skilled sectors such as banking and law, governments may need to start considering a form of care-oriented economy based on a universal credit system overseen by central banks. Lindgren suggests that this may become necessary to avoid the marginalisation of large swathes of populations who can no longer find work in profit-making institutions. He refers to an example of a society where an energy transformation to small-scale cheap renewable energy (Lindgren actually uses the example of mini nuclear reactors) affords universal access to low-cost energy and subsequent proliferation of automation replacing the need for human labour across most sectors of the economy. This could potentially free people's time for care-oriented activities either in self-care, community care, or in care for the planet, in return for access to a universal income from a more firmly central bank-controlled economic system (Lindgren, 2024).

Whatever the drivers of a transformation, witnessing the state of play on the global stage at the time of writing of this book in early 2026 suggests that a transformation of sorts is already in motion. Sustainable development may not be the direction favoured by the current ruling elite – the leaders with the power reputation – but there is hope yet in integrated governance approaches that build legitimacy through reflexive practice, offer means to resist the dominant trajectory of development and pioneer and advocate for sustainable and transformational solutions that serve just and regenerative outcomes.

8.5 Conclusions

Governance interventions that aspire to further a societal transformation that addresses the normative and systemic drivers of unsustainability must avoid simplistic incremental fixes. Many so-called sustainable development solutions aim to maintain the status quo in terms of power and existing inequalities, sometimes further enhancing them. Solutions that work in favour of the dominant power 'elites' such as strong business interests and populist politicians will only delay the kind of change that is required for humanity to survive the ongoing planetary crises. It is extremely important that governance efforts remain ambitious and resistant to the efforts by powerful interests to distract from the task at hand. Governance must acknowledge and respect the complexity of the 21st-century context and the challenges posed by the prevalent crises and entrenched extractive perspectives, values and practices.

If power is understood as a system capacity rather than something that can be held by an individual or an organisation alone, then it makes sense that a deliberative approach fostering plurality and learning needs to permeate governance (Voß & Kemp, 2006). Chapter 7 of this volume summarises the principles of good governance in four key tenets. The first principle pertains to the scope of governance, as *planetary*, and attendant attention to care and regeneration of the planetary system that includes but does not prioritise humans. The planetary focus decentres humans as the key focus of governance and highlights our dependence on both non-human and non-living environments that have a right to exist and thrive, if not for any other reason than that we depend on them (San Martín & Wood, 2021). The appreciation of the Earth as a complex system, where the interdependencies between different domains of the environment and the economy need to be acknowledged and considered, stipulates that governance efforts need to be *integrated* – involving coordination across different administrative and disciplinary sectors, formal and informal stakeholders, and governance and territorial scales. In the context of a hugely unequal planet mired in multiple crises, all governance efforts need to pay attention to *legitimacy* – that is, they need to serve just ends in a fair manner and need to be acceptable, perhaps as compromises, to the broad range of interests and perspectives that they implicate, from the human to the non-human, from Western to Indigenous and from the present day to the future (Hiedanpaa & Bromley, 2016). *Reflexivity*, the final principle, mainly pertains to the manner in which governance is undertaken. It stipulates a focus on learning by all stakeholders, formal and informal, throughout, reliance on plural knowledge, open discussion of conflicting values and interests, and flexibility, the ability to adapt following feedback from ongoing evaluation (Voß & Kemp, 2006).

The principles respond to the challenges of governing climate change and food security set out in Chapters 2, 5, and 6 of this volume and provide an ideal rather than a blueprint. The discussion of evaluation of both governance processes and outcomes in Chapter 7 is designed to support readers in beginning to apply the ideas set out in this book into practice in proposing and evaluating governance interventions for sustainable development in the 21st century. There is no straightforward formula for getting governance right in reality and a place-sensitive approach remains hugely important and indeed necessary for the realisation of the above four principles. The most pressing task for those who aspire to govern for transformational sustainable development remains to inspire the political will of formal governance actors and the willingness of stakeholders and communities to engage in governance as ongoing efforts to steer decision-making towards solutions that serve the wellbeing of the planet and the ability of all its inhabitants to thrive.

References

Anderson, K. (2023, March 24). IPCC's conservative nature masks true scale of action needed to avert catastrophic climate change. *The Conversation*. Retrieved September 5, 2025, from http://theconversation.com/ipccs-conservative-nature-masks-true-scale-of-action-needed-to-avert-catastrophic-climate-change-202287

Beck, U. (2006). Reflexive governance: Politics in the global risk society. In *Reflexive governance for sustainable development*. Edward Elgar Publishing. https://doi.org/10.4337/9781847200266.00011

Biermann, F., Hickmann, T., Sénit, C. A., Beisheim, M., Bernstein, S., Chasek, P., Grob, L., Kim, R. E., Kotzé, L. J., Nilsson, M., Ordóñez Llanos, A., Okereke, C., Pradhan, P., Raven, R., Sun, Y., Vijge, M. J., van Vuuren, D., Wicke, B., Global Sustainability Governance, ... Energy and Resources. (2022). Scientific evidence on the political impact of the sustainable development goals. *Nature Sustainability*, *5*(9). https://doi.org/10.1038/s41893-022-00909-5

Boswell, J., Dean, R., & Smith, G. (2022). Integrating citizen deliberation into climate governance: Lessons on robust design from six climate assemblies. *Public Administration*, *101*(1), 182. https://doi.org/10.1111/padm.12883

Bulkeley, H., & Mol, A. P. J. (2003). Participation and environmental governance: Consensus, ambivalence and debate. *Environmental Values*, *12*(2), 143–154. https://doi.org/10.3197/096327103129341261

Chancel, L., Gómez-Carrera, R., Moshrif, R., & Piketty, T. (2026). *World inequality report 2026*. https://wid.world/document/world-inequality-report-2026/

Cherry, C., Capstick, S., Demski, C., Mellier, C., Stone, L., & Verfuerth, C. (2021). *Citizens' climate assemblies: Understanding public deliberation for climate policy*. Cardiff University. https://orca.cardiff.ac.uk/id/eprint/145771/

Cohen, J., Dannereuther, C., Fraundorfer, M., Macie, C., Martin-Ortega, J., Mdee, A., & Salazar Sutin, N. (2023). Riverkin: Seizing the moment to remake vital relations in the United Kingdom and beyond. *People and Nature*, *5*, 1877–1892. https://doi.org/10.1002/pan3.10534

Coolsaet, B. (Ed.). (2020). *Environmental justice: Key issues*. Taylor & Francis Group.

Douwe Van Der Ploeg, J. (2013). Peasant-driven agricultural growth and food sovereignty. *The Journal of Peasant Studies*, *40*(6), 999–1030.

Driscoll, D. (2023). Populism and carbon tax justice: The yellow vest movement in France. *Social Problems*, *70*(1), 143–163. https://doi.org/10.1093/socpro/spab036

Extinction Rebellion. (2025). *Citizens' assembly*. Extinction Rebellion UK. Retrieved January 20, 2026, from https://extinctionrebellion.uk/decide-together/citizens-assembly/

Folke, C., Polasky, S., Rockström, J., Galaz, V., Westley, F., Lamont, M., Scheffer, M., Österblom, H., Carpenter, S. R., Chapin, F. S., Seto, K. C., Weber, E. U., Crona, B. I., Daily, G. C., Dasgupta, P., Gaffney, O., Gordon, L. J., Hoff, H., Levin, S. A., ... Walker, B. H. (2021). Our future in the Anthropocene biosphere. *Ambio*, *50*(4), 834–869. https://doi.org/10.1007/s13280-021-01544-8

Gallent, N., Hamiduddin, I., Juntti, M., Kidd, S., & Shaw, D. (2015). *Introduction to rural planning: Economies, communities and landscapes* (2nd ed.). Routledge. https://doi.org/10.4324/9781315749280
Grant, W. (2021). *¡Populista! The rise of Latin America's 21st century strongman.* Head of Zeus Ltd.
Grossman, E. (2020). *Emergency on planet earth. Environmental Health Perspectives, 128*(5), 054001.
Gupta, J., Liverman, D., Prodani, K., Aldunce, P., Bai, X., Broadgate, W., Ciobanu, D., Gifford, L., Gordon, C., Hurlbert, M., Inoue, C. Y. A., Jacobson, L., Kanie, N., Lade, S. J., Lenton, T. M., Obura, D., Okereke, C., Otto, I. M., Pereira, L., ... Verburg, P. H. (2023). Earth system justice needed to identify and live within Earth system boundaries. *Nature Sustainability*, 6(6). https://doi.org/10.1038/s41893-023-01064-1
Hiedanpaa, J., & Bromley, D. (2016). *Environmental heresies: The quest for reasonable.* Palgrave Macmillan. https://doi.org/10.1057/978-1-137-60083-7
IPCC. (2023). *Climate change 2023: Synthesis report. Contribution of working groups I, II and III to the sixth assessment report of the intergovernmental panel on climate change* [Core Writing Team, H. Lee & J. Romero, Eds.). IPCC. https://www.ipcc.ch/report/ar6/syr/
Jackson, T. (2017). *Prosperity without growth: Foundations for the economy of tomorrow* (2nd ed.). Routledge. https://doi.org/10.4324/9781315677453
Jackson, P., Guadalupe, M., Ferre, R., Candel, J., Davies, A., Derani, C., De Vries, H., Dragović-Uzelac, V., Hoel, A. H., Holm, L., Mathijs, E., Morone, P., Penker, M., Śpiewak, R., Termeer, K., & Thøgersen, J. (2021). *Food as a commodity, human right or common good Different framings of food may shape food policies and their impact. Despite acknowledging food systems' complexities, the European Commission's Farm to Fork Strategy still addresses food as a commodity instead of a human right or common good. Nature Food*, 2(3), 132–134.
Jackson, T. (2025). *The care economy.* Polity.
Jones, M., & Hill, S. (2023). *Local food partnerships in wales sustainable food places evaluation report.* UWE Bristol. https://uwe-repository.worktribe.com/OutputFile/10994402
Liegey, V., & Nelson, A. (2020). *Exploring Degrowth: A critical guide* (1st ed.). Pluto Press.
Lindgren, J. (2024). *Ohjaustalous vai nakymaton kasi.* Vastapaino.
Marsden, T., Faus, A., & Sonnino, R. (2019). Reproducing vulnerabilities in agri-food systems: Tracing the links between governance, financialization, and vulnerability in Europe post 2007–2008. *Agrarian Change.* https://doi.org/10.1111/joac.12267
Our Food. (2025). *Our food local, profitable, sustainable.* Retrieved January 20, 2026, from https://our-food.org/
Oxfam. (2023). *Climate equality: A planet for the 99%.* Oxfam GB. https://doi.org/10.21201/2023.000001
Redclift, M. (2005). Sustainable development (1987–2005): An oxymoron comes of age. *Sustainable Development, 13*, 212–217. https://doi.org/10.1002/sd.281
Ritchie, H. (2021). *Smallholders produce one third of world's food, less than half of what many headlines claim.* Our World In Data. https://ourworldindata.org/smallholder-food-production
Ritchie, H., & Roser, M. (2024, January 22). *CO_2 emissions. How much CO_2 does the world emit? Which countries emit the most?* Our World in Data. Retrieved March 12, 2024, from https://ourworldindata.org/co$_2$-emissions
San Martín, W., & Wood, N. (2021). Pluralising planetary justice beyond the North-South divide: Recentring procedural, epistemic, and recognition-based justice in

earth-systems governance. *Environmental Science & Policy*, *128*, 256. https://doi.org/10.1016/j.envsci.2021.12.002

Scott, M., Parker, G., Juntti, M., Castellino, J., Forero, O., Mell, I., Jerome, G., Amati, M., Hernandez, C., Buntine, C., Dodd, A., Hamiduddin, I., Brien, C. O., Lucocq, H., Lennon, M., Blyth, R., Parker, G., & Scott, M. (2024). The biodiversity crisis – planning for nature recovery? *Planning Theory & Practice*, *25*(1), 103–140. https://doi.org/10.1080/14649357.2024.2322879

Scott, M., Sturzaker, J., & Juntti, M. (2025). The 'economic rural' – supporting recovery and regeneration. In M. Scott, J. Sturzaker, N. Gallent, G. Parker, A. Burnett, & I. Mell (Eds.), *Rural planning futures*. Routledge. https://doi.org/10.4324/9781003423164

Steffen, W., Richardson, K., Johan Rockström, Cornell, S. E., Fetzer, I., Bennett, E. M., Biggs, R., Carpenter, S. R., Wim, V., Wit, C. A., Folke, C., Gerten, D., Heinke, J., Mace, G. M., Persson, L. M., Ramanathan, V., Reyers, B., & Sörlin, S. (2015). Planetary boundaries: Guiding human development on a changing planet. *Science*, *347*(6223), 1259855. https://doi.org/10.1126/science.1259855

Sustainable Food Places. (2025). *Monmouthshire | Sustainable food places*. Retrieved Jan 20, 2026, from https://www.sustainablefoodplaces.org/members/monmouthshire/

UN General Assembly. (2015). *Transforming our world: The 2030 Agenda for Sustainable Development* (A/RES/70/1). United Nations. Avaiable online at: https://digitallibrary.un.org/record/3923923?v=pdf

Voß, J. P., & Kemp, R. (2006). Sustainability and reflexive government: Introduction. In *Reflexive governance for sustainable development*. Edward Elgar Publishing. https://doi.org/10.4337/9781847200266.00009

Index

Agri-environmental policy (AEP) 118
Aldaya, M.M. 119
Allen, J. 63
Anderson, K. 90
Anthropocene 22, 133

Bailey, I. 95
Beck, U. 20, 22, 30, 89–90
Biermann, F. 95
Boswell, J. 153, 155

Carbon equivalent 85
Carbon pricing 143
Care workers 159
Castells, M. 28–29
Chancel, L. 19, 28
Cherry, C. 154
Chirambo, D. 48
Cities for Climate Protection (CCP) Programme 96, 136
Clark, M. 112–113
Clean Development Mechanism (CDM) 93
Clean Energy Mechanism (CEM) 95
Climate change 83; adaptation as transformation 97; adaptation as transition 97; adaptation goals 97; adaptation policy 96; agricultural productivity 111; aspirations 92; dimensions of justice 98; GHG emissions 98; global governance 91–92; mitigation of emissions 112; mitigation targets 92–93, 95; social and cultural constructions 84; socio-environmental issues 84; technological fixes 100
Climate change entrepreneurship 48–49
Climate crisis 8, 20
Climate justice 84, 88, 98, 132, 151
CO_{2e} emissions 85–86
Cohen, J. 155
Colonial-era power 99
Commodity values 27
Common Agricultural Policy (CAP) 116–117; AEP 118; EU Agri-environmental policy 118; family-farm structure 118; objectives of 117
Conceptual learning 140
Conference of Parties (COP) 70, 84
Constructivism 68
Convention Citoyenne pour le Climat (CCC) 154
Courgette crisis 23
Crisis 17; definition of 8, 17; income and wealth 18–19; socio-economic dimension 18

Derivatives 27
Developed countries 26, 37, 68, 88, 131–132
Developed economies 91, 160
Digitised centralised trading 27
Digitised commodity networks 28
Distributional justice 46, 115

Earth system boundaries 22–23, 50–51, 133
Earth system governance 73
Ecological ceiling 51
Ecosystem services 42
Environmental justice 47
Environmental sustainability 4, 119
Epistemic colonialism 19
EU institutions 72
Extinction Rebellion (XR) 7, 153

Fact checking 19
Financialisation 110
Food deserts 108
Food insecurity 106–107

Food miles 113
Food security 104; access 105; agency 105; availability 105; childhood obesity 109; definition 104; dietary change 109; environmental sustainability 106; international food system 105, 116–117; livelihoods change 108; pillars of 104; sustainability 105; trade agreements 106; utilisation 105
Food sovereignty 114, 116–117, 121, 151
Food system 26, 106; agency of consumers 119; agricultural policies 117; agri-tech companies 120; agro-ecological practices 113; business structures 114; complexity 156; cultural value 114; distirbutive justice 115; environmental sustainability 119, 155; frugal abundance 159; governance of trade 118–119; holistic sustainability agenda 121; local food 120; organic production 113; procedural justice 115; production-consumption relationship 113; productivist 109; productivist liberal agenda 121; product labels 119; public schemes 119; quantity of food 116–117; relational power 155; right to food 115–116; sectoral policies 116; sustainable food production 156; UK 156–158
Formal governance 5, 155
Formal rationality 75
Fossil capitalism 90
Futures 27

Genetic diversity 21
Genetic modification technologies 111
Genuine Development Index (GDI) 38
Ghosh, A. 30
Gig 28
Glasgow consensus 95
Global governance 6–7, 61–62, 116
Globalisation 23
Global warming 20
Goal-achievement model 141
Goal-free evaluation model 141–142
Good governance 7, 74
Governance 5; in 21st century 17; agenda setting 139; bottom-up approach 65; challenge 29; Climate and Resilience Bill 154; constructivist approach 67–68; conventional evaluation 140; cost-benefit 143; crisis 7–8; cultural norms 63; cyclical process 61; decision-making 60, 139, 141; definition 5; democratic institutions 60; design phase 60; environmental actors 62; environmental policies 142; ex ante evaluation 139–140; ex post evaluation 141; flexibility 143; formal actors 61; formal/informal actors 62; France 154; industrial revolution 99; informal actors 61; integration 139; international regimes 69; international scale 6–7; interventions 64; issues 7; multilevel 71–73; national scale 6–7; networked 65; online sessions 154; operative level 65; participation 140; policy sectors 140; polycentric 65; power 60, 62–63; power of technologies 63–64; qualitative categorisation 142; recipients of 63; services and goods 6; social media 30; systematic evaluation 141–142; top-down approach 64–65; types of knowledge 65; usable knowledge 62
Governance fragmentation 73
Governance institutions 60–61
Governance outcomes 61
Governance outputs 61
Green Climate Fund 93
Green economy 59, 96
Greenhouse gas (GHG) emissions 20, 85, 92, 111
Green tax 60
Gross Domestic Product (GDP) 38
Growth-based economic system 91
Gupta, J. 136

Haberl, H. 48–49
Hickel, J. 51
High Level Panel of Experts on Food Security (HLPE) 104, 116–117
Hooghe, L. 71
Hunger 108, 117

Industrial food system 114–115
Industrialisation 20, 22
Inequality crisis 19, 29–30
Informal governance 119–120
Intergovernmental Panel on Climate Change (IPCC) 20

Intergovernmental Science-Policy Platform on Ecosystem Services (IPBES) 21
International Energy Agency (IEA) 86
International food system 26, 109, 116–117
Internationalisation 110
International Panel on Climate Change (IPCC) 85

Jackson, T. 43, 159
Jordan, A. 140
Just Energy Transition Partnerships (JETPs) 89, 94

Kim, R. 95
Kirwan, J. 120
Kothari, A. 51
Kyoto Protocol 93

Labour 24
Legitimacy 60, 161
Liegey, V. 53, 159
Lindgren, J. 160
Lobbying 72
Local food 113, 120
Long Term Emission Development Strategies (LT-EDS) 93
Lukes, S. 62

Market logic 6
Marks, G. 71
Marsden, T. 109
Maye, D. 120
McDermott, T.K.J. 97
Mickwitz, P. 141–142
Millennium Ecosystem Assessment (MEA) 43
Mini publics 153
Mitigation burden 89
Modernisation 20, 38
Modernity 22
Monmouthshire Food Partnership 157
Multilevel governance approach 71–73
Multinational Corporations (MNCs) 68
Multiple crises 8

Nationally Determined Contributions (NDCs) 99, 152
Nature restoration 43
Nelson, A. 53, 159
Neoliberalism 5
Net Inventory Reports (NIRs) 92, 152
Network governance 66
Network society theory 24, 27–28
New economic policy 6
Non-decision-making 62
Non-Governmental Organisations (NGOs) 68, 75

Organisation for Economic Cooperation and Development (OECD) 68

Paris Agreement 99
Piketty, T. 25
Planetary crisis 17, 20, 23, 127; modern truths 31; poverty 45; risk 29; societal implications 22
Platform economy 28
Populism 132, 153
Post-Growth Conference 91
Poverty 37
Prevalence of undernourishment (PoU) 108
Proactive governance 119–120
Procedural justice 46, 115
Public expenditure 25
Public goods 61
Public interest 6
Public-private partnerships 72

Raworth, K. 50–51
Reflexive governance 75, 137
Relative decoupling 48
Resilience 22, 97
Restitution payments 89
Restorative justice 89
Rhodes, R.A.W. 66
Richie, H. 87
Rosenthal, U.E. 8
Russel, D. 140

Scott, M. 157
Side-effects evaluation model 141–142
Social media 19
Socio-economic system 24
Street level bureaucrats 75
Super smart society 49
Surminski, S. 97
Sustainability skills gap 50
Sustainable development 3, *4*, 127; 21st-century crises 74; climate change 41; climate change 42; complexity 130; crisis of knowledge 132; definition 3, 36; definition 59; dependence on nature 46–47;

distributional justice 46; Earth system boundaries 133–134; economy 38; environment 38–39; environmental degradation 36; environmental injustices 132; environmental justice 47; extractive processes 159; food system 115; global governance 68; Global South 36; governance ambition and implementation 131–132; green economy 48–50; growth imperative 129–130; implementation gap 39; legitimacy 135–137; market economy 158, 160; moral dilemmas 60; non-hierarchical modes 158; people and communities 45; planetary system 39; policy domains 134–135; procedural justice 46; recognition justice 46; reflexive governance 137–138; scientific knowledge 132–133; social dimension 46; society 38; socio-economic inequality 45; socio-technical systems 130–131; system boundaries 39; technology 160; transformation 50–53; transformational sustainability 128
Sustainable food system 112, 117
Sustainable productivism 112–113, 117
Sze, J. 8

Taxation 25
Thomas, C. 97
Tilman, D. 112–113

United Nations Framework Convention on Climate Change (UNFCCC) 83–84

Virtual water 119
Voulvoulis, N. 130

Weiss, T.G. 74
Wilkinson, R. 74
World Commission on Environment and Development (WCED) 36

Yellow vest protests 60, 153

For Product Safety Concerns and Information please contact our EU representative GPSR@taylorandfrancis.com
Taylor & Francis Verlag GmbH, Kaufingerstraße 24, 80331 München, Germany

www.ingramcontent.com/pod-product-compliance
Ingram Content Group UK Ltd.
Pitfield, Milton Keynes, MK11 3LW, UK
UKHW020316270926
475702UK00050B/800

* 9 7 8 1 0 3 2 0 0 5 9 9 7 *